THE OFFICIAL CIA INTERROGATION- TION AND MANIPULATION MANUAL

THE OFFICIAL CIA INTERRO-GATION & MANIPULATION MANUAL

THE COLD WAR KUBARK FILES

ISBN: 978-1-963956-23-8

Contents

I. INTRODUCTION

A. Explanation of Purpose

This manual cannot teach anyone how to be, or become, a good interrogator. At best it can help readers to avoid the characteristic mistakes of poor interrogators. Its purpose is to provide guidelines for KUBARK interrogation, and particularly the counterintelligence interrogation of resistant sources. Designed as an aid for interrogators and others immediately concerned, it is based largely upon the published results of extensive research, including scientific inquiries conducted by specialists in closely related subjects. There is nothing mysterious about interrogation. It consists of no more than obtaining needed information through responses to questions.

As is true of all craftsmen, some interrogators are more able than others; and some of their superiority may be innate. But sound interrogation nevertheless rests upon a knowledge of the subject matter and on certain broad principles, chiefly psychological, which are not hard to understand. The success of good interrogators depends in large measure upon their use, conscious or not, of these principles and of processes and techniques deriving from them. Knowledge of subject matter and of the basic principles will not of itself create a successful interrogation, but it will make possible the avoidance of mistakes that are characteristic of poor interrogation. The purpose, then, is not to teach the reader how to be a good interrogator but rather to tell him 4 what he must learn in order to become a good interrogator.

The interrogation of a resistant source who is a staff or agent member of an Orbit intelligence or security service or of a clandestine Communist organization is one of the most exacting of professional tasks. Usually the odds still favor the interrogator, but they are sharply cut by the training, experience, patience and toughness of the interrogatee. In such circumstances the interrogator needs all the help that he can get. And a principal source of aid today is scientific findings. The intelligence service which is able to bring pertinent, modern knowledge to bear upon its problems enjoys huge advantages over a service which conducts its clandestine business in eighteenth century fashion. It is true that American psychologists have devoted somewhat mote attention to Communist interrogation techniques, particularly "brainwashing", than to U.S. practices. Yet they have conducted scientific inquiries into many subjects that are closely related to interrogation: the effects of debility and isolation, the polygraph, reactions to pain and fear, hypnosis and heightened suggestibility, narcosis, etc. This work is of sufficient importance and relevance that it is no longer possible to discuss interrogation significantly without reference to the psychological research conducted in the past decade. For this reason a major purpose of this study is to focus relevant scientific findings upon CI interrogation. Every effort has been made to report and interpret these findings in our own language, in place of the terminology employed by the psychologists. This study is by no means confined to a resume and interpretation of psychological findings. The approach of the psychologists is customarily manipulative; that is, they suggest methods of imposing controls or alterations upon the interrogatee from the outside. Except within the Communist frame of reference, they have paid less attention. to the creation of internal controls--i.e., conversion of the source, so that voluntary cooperation results. Moral considerations aside, the imposition of external techniques of manipulating people carries with it the grave risk of later lawsuits, adverse publicity, or other attempts to strike back.

B. Explanation of Organization

This study moves from the general topic of interrogation per se (Parts I, I, II, IV, V, and VI) to planning the counterintelligence interrogation (Part VII) to the Cl interrogation of resistant sources (Parts VII, IX, and X). The definitions, legal considerations, and discussions of interrogators and sources, as well as Section VI on screening and other preliminaries, are relevant to all kinds of interrogations. Once it is established that the source is probably a counterintelligence target (in other words, is probably a member of a foreign intelligence or security service, a Communist, or a part of any other group engaged in clandestine activity directed against the national security), the interrogation is planned and conducted accordingly. The CI interrogation techniques are discussed in an order of increasing intensity as the focus on source resistance grows sharper. The last section, on do's and dont's, is a return to the broader view of the opening parts; as a check-list, it is placed last solely for convenience.

II. DEFINITIONS

Most of the intelligence terminology employed here which may once have been ambiguous has been clarified through usage or through KUBARK instructions. For this reason definitions have been omitted for such terms as burn notice, defector, escapee, and refugee. Other definitions have been included despite a common agreement about meaning if the significance is shaded by the context.

1. Assessment: the analysis and synthesis of information, usually about a person or persons, for the purpose of appraisal. The assessment of individuals is based upon the compilation and use of psychological as well as biographical detail.

2. Bona fides: evidence or reliable information about identity, personal (including intelligence) history, and intentions or good faith.

3. Control: the capacity to generate, alter, or halt human behavior by implying, citing, or using physical or psychological means to ensure compliance with direction. The compliance may be voluntary or involuntary. Control of an interrogatee can rarely be established without control of his environment,

4. Counterintelligence interrogation: an interrogation (see #7) designed to obtain information about hostile clandestine activities and persons or groups engaged therein. KUBARK CI interrogations are designed, almost invariably, to yield information about foreign intelligence and security services or Communist organizations. Because security is an element of counterintelligence, interrogations conducted to obtain admissions of clandestine plans or activities directed against KUBARK or PBPRIME security are also CI interrogations. But unlike a police interrogation, the CI interrogation is not aimed at causing the interrogatee to incriminate himself as a means of bringing him to trial. Admissions of complicity are not, to a CI service, ends; in themselves but merely preludes to the acquisition of more information.

5. Debriefing: obtaining information by questioning a controlled and witting source who is normally a willing one.

6. Eliciting: obtaining information, without revealing intent or exceptional interest, through a verbal or written exchange with a person who may be willing or unwilling to provide what is sought and who may or may not be controlled.

7. Interrogation: obtaining information by direct questioning

of a person or persons under conditions which are either partly or fully controlled by the questioner or are believed by those questioned to be subject to his control. Because interviewing, debriefing, and eliciting are simpler methods of obtaining information from cooperative subjects, interrogation is usually reserved for sources who are suspect, resistant, or both. 8. Intelligence interview: obtaining information, not customarily under controlled conditions, by questioning a person who is aware of the nature and perhaps of the significance of his answers but who is ordinarily unaware of the purposes and specific intelligence affiliations of the interviewer.

III. LEGAL AND POLICY CONSIDERATIONS

The legislation which founded KUBARK specifically denied it any law-enforcement or police powers.| Yet detention in a controlled environment and perhaps for a lengthy period is frequently essential to a successful counterintelligence interrogation of a recalcitrant source. [Because the necessary powers are vested in the competent liaison service or services, not in KUBARK, it is frequently necessary to conduct such interrogations with or through liaison"] This necessity, obviously, should be determined as early as possible. The legality of detaining and questioning a person, and of the methods employed, is determined by the laws of the country in which the act occurs. [If;is therefore important that all KUBARK a interrogators and their supervisors be fully and accurately informed about the applicable local laws. This principle holds whether the interrogation is to be conducted unilaterally or jointly. It is unsafe to assume that all members of the liaison service know the pertinent statutes.

Moreover, a joint illegal interrogation may later embarrass both services and lead to recriminations and strained relations between them. It is recommended that copies or legal extracts of all applicable laws be kept by the Station or Base in a separate file and that all concerned reread the file per logically. Detention poses the most common of the legal problems. KUBARK has no independent legal authority to detain anyone against his will, (and liaison services may not, as a rule, legally confer such authority upon KUBARK.

Even if the local authorities have exercised powers of detention in our behalf, the legal time-limit may be narrow. The haste in which some KUBARK interrogations have been conducted has not always been the product of impatience. Some security services, especially those of the Sino-Soviet Bloc, may work at leisure, depending upon time as well as their own methods to melt recalcitrance. KUBARK usually cannot. Accordingly, unless it is considered that the prospective interrogatee is cooperative and will remain so indefinitely, the first step in planning an interrogation is to determine how long the source can be held. The choice of methods i dena in part upon the answer to this question.

The questioning of defectors are subject to the provisions of directive No. 4; to its related Chief/KUBARK ie) Directives, principal Book Dispatch (b)(3) and to pertinent(Those concerned with the (b)(3) "interrogation of defectors, escapees, refugees, or repatriates should know these references. : The kinds of counterintelligence information to be sought in a CI interrogation are stated generally in Chief RK Directive and in greater detail in Book Dispatch (b) (3) The interrogation of PBPRIME citizens poses special problems. First, such interrogations should not be conducted for reasons lying outside the sphere of KUBARK's responsibilities. For example, the security of other ODYOKE departments and agencies overseas is their : own responsibility. KUBARK may provide behind-the-scenes assistance-for example, (b)(1) 'but should not normally (b)(3) "become directly involved. Clandestine activity conducted abroad on behalf of a foreign power by a private PBPRIME citizen does fall within KUBARK's investigative and interrogative responsibilities. However, any investigation, interrogation, or interview of a PBPRIME citizen which is conducted abroad because it is known or suspected that he is engaged in clandestine activities directed against PBPRIME security interests requires the prior and personal approval of Chief/KUDESK or of his deputy.

Since 4 October 1961, extraterritorial application has been given to the Espionage Act, making it henceforth possible to prosecute in the Federal Courts any PBPRIME citizen who violates the statutes of this Act in foreign countries. ODENVY has requested that it be informed, in advance if time permits, if any investigative steps are undertaken in these cases. Since KUBARK employees cannot be witnesses in court, each investigation must be conducted in such a manner that evidence obtained may be properly introduced if the case comes to trial. States policy and procedures for the conduct of investigations of PBPRIME citizens abroad. Interrogations conducted under compulsion or duress are especially likely to involve illegality and to entail damaging consequences for KUBARK.

Therefore prior Headquarters approval at the KUDOVE level must be obtained for the interrogation of any source against his will and under any of the following circumstances: 1. If bodily harm is to be inflicted. 2. If medical, chemical, or electrical methods or materials are to be used to induce acquiescence. 3. If the detention is locally illegal and traceable to KUBARK, except that in cases of extreme operational urgency requiring immediate detention, retroactive Headquarters approval may be promptly requested by priority cable. The CI interrogator dealing with an uncooperative interrogatee who has been well-briefed by a hostile service on the legal restrictions under which ODYOKE services operate must expect some effective delaying tactics. The interrogatee has been told that KUBARK will not hold him long, that he need only resist for a while. Nikolay KHOKHLOV, for example, reported that before he left for Frankfurt am Main on his assassination mission, the following thoughts coursed through his head: "If I should get into the hands of Western authorities, I can become reticent, silent, and deny my voluntary visit to Okolovich. I know I will not be tortured and that under the procedures of western law I can conduct myself boldly." (17) [1 The footnote numerals in this text are keyed to the numbered bibliography at the end. / The interrogator who encounters expert resistance should not grow flurried and press; if he does, he is likelier to commit illegal acts which the source can later use against him. Remembering that time is on his side, the interrogator should arrange to get as much of it as he needs,

IV. THE INTERROGATOR

A number of studies of interrogation discuss qualities said to be desirable in an interrogator. The list seems almost endless a professional manner, forcefulness, understanding and sympathy, breadth of general knowledge, area knowledge, "a practical knowledge of psychology", skill in the tricks of the trade, alertness, perseverance, integrity, discretion, patience, a high IQ, extensive experience, flexibility, etc., etc. Some texts even discuss the interrogator's manners and grooming, and one prescribed the traits considered desirable in his secretary. A repetition of this catalogue would serve no purpose here, especially because almost all of the characteristics mentioned are also desirable in case officers, agents, policemen, salesmen, lumberjacks, and everybody else. The search of the pertinent scientific literature disclosed no reports of studies based on common denominator traits of successful interrogators or any other controlled inquiries that would invest these lists with any objective validity.

Perhaps the four qualifications of chief importance to the interrogator are (1) enough operational training and experience to permit quick recognition of leads; (2) real familiarity with the language to be used; (3) extensive background knowledge about the interrogatee's native country (and intelligence service, if employed by one); and (4) a genuine understanding of the source as a person. K defector center, some Stations, and even a few bases can call upon one or several interrogators to supply these prerequisites, individually or as a team. Whenever a number of interrogators is available, the percentage of successes is increased by careful matching of questioners and sources and by ensuring that rigid prescheduling does not prevent such matching. Of the four traits listed, a genuine insight into the source's character and motives is perhaps most important but least common. Later portions of this manual explore this topic in more detail. One general observation is introduced now, however, because it is considered basic to the establishment of rapport, upon which the success of non-coercive interrogation depends.

The interrogator should remember that he and the interrogatee are often working at cross-purposes not because the interrogatee is malevolently withholding or misleading but simply because what he wants from the situation is not what the interrogator wants. The interrogator's goal is to obtain useful information-facts about which the interrogatee presumably has acquired information. But at the outset of the interrogation, and perhaps for a long time afterwards, the person being questioned is not greatly concerned with communicating his body of specialized information to his questioner; he is concerned with putting his best foot forward. The question uppermost in his mind, at the beginning, is not likely to be "How can I help PBPRIME?" but rather "What sort of impression am I making?" and, almost immediately thereafter, "What is going to happen to me now?" (An exception is the penetration agent or provocateur sent to a KUBARK field installation after training in withstanding interrogation. Such an agent may feel confident enough not to be gravely concerned about himself. His primary interest, from the beginning, may be the acquisition of information about the interrogator and his service.) The skilled interrogator can save a great deal of time by understanding the emotional needs of the interrogatee. Most people confronted by an official--and dimly powerful--representative of a foreign power will get down to cases much faster if made to feel, from the start, that they are being treated as individuals. So simple a matter as greeting an interrogatee by his name at the opening of the session establishes in his mind the comforting awareness that he is considered as a person, not a squeezable sponge. This is not to say that egotistic - types should be allowed to bask at length in the warmth of individual recognition.

But it is important to assuage the fear of denigration which afflicts many people when first interrogated by making it clear that the individuality of the interrogatee is recognized. With this common understanding established, the interrogation can move on to impersonal matters and will not later be thwarted or interrupted-uu seghet or at least not as often--by irrelevant answers designed not to provide facts but to prove that the interrogatee is a respectable member of the human race. Although it is often necessary to trick people into telling what we need to know, especially in Cl interrogations, the initial question which the interrogator asks of himself should be, "How can I make him want to tell me what he knows?" rather than "How can I trap him into disclosing what he knows?" If the person being questioned is genuinely hostile for ideological reasons, techniques of manipulation are in order. But the assumption of hostility--or at least the use of pressure tactics at the first encounter--may make difficult subjects even out of those who would respond to recognition of individuality and an initial assumption of good will. Another preliminary comment about the interrogator is that z normally he should not personalize. That is, he should not be as pleased, flattered, frustrated, goaded, or otherwise emotionally ne and personally affected by the interrogation. A calculated display = of feeling employed for a specific purpose is an exception; but even under these circumstances the interrogator is in full control. The interrogation situation is intensely inter-personal; it is therefore all the more necessary to strike a counter-balance by an attitude which the subject clearly recognizes as essentially fair and objective. The kind of person who cannot help personalizing, who becomes emotionally involved in the interrogation situation, may have chance (and even spectacular) successes as an interrogator but is almost certain to have a poor batting average. It is frequently said that the interrogator should be "a good judge of human nature." In fact, "all interrogation guides stress that ts is important to 'size up the source's personality'; yet research can show little reliability or validity in the evaluations which are made in such circumstances.'" (3) This study states later (page "Great attention has been given to the degree to which persons are ; able to make judgements from casual observations regarding the ' personality characteristics of another.

The consensus of research is that with respect to many kinds of judgments, at least some judges 'perform reliably better than chance....' Nevertheless, "...the level of reliability in judgments is so low that research encounters difficulties when it seeks to determine who makes better judgments... ." (3) In brief, the interrogator is likelier to overestimate his ability to judge others than te underestimate it, especially if he has had little or no training in modern psychology. It follows that errors in assessment and in handling are likelier to result from snap judgments based upon the assumption of innate skill in judging others than from holding such judgments in abeyance until enough facts are known. There has been a good deal of discussion of interrogation experts vs. subject-matter experts. Such facts as are available suggest that the latter have a slight advantage. But for counterintelligence purposes the debate is academic. [The CI interrogator must be both highly knowledgeable about the hostile service, CP, or other group with which the interrogatee may be linked* and highly skillful in the art of interrogation. If a man who has both kinds of knowledge is not available when the CI interrogation must be conducted, it is better to use a two-man team, each interrogator supplementing the other. It is sound practice to assign inexperienced interrogators to guard duty or to other supplementary tasks directly related to interrogation, so that they can view the process closely before taking charge. The use of beginning interrogators as screeners (see part VI) is also recommended. Although there is some limited validity in the view, frequently expressed in interrogation primers, that the interrogation is essentially a battle of wits, the CI interrogator who encounters a skilled and resistant interrogatee should remember that a wide *The interrogator should be supported whenever possible by ' qualified analysts' review of his daily "take"; experience has shown that such a review will raise questions to be put and points to be clarified and lead to a thorough coverage of the subject in hand at variety of aids can be made available in the field or from Headquarters. (These are discussed in Part VIII.) The intensely personal nature of the interrogation situation makes it all the more necessary that the KUBARK questioner should aim not for a personal triumph but for his true goal--the acquisition of all needed information by any authorized means.

V. THE INTERROGATEE

A. Types Of Sources: Intelligence Categories

From the viewpoint of the intelligence service the categories of persons who most frequently provide useful information in response to questioning are travellers; repatriates; defectors, escapees, and refugees; transferred sources; agents, including provocateurs, double agents, and penetration agents; and swindlers and fabricators.

1. Travellers are usually interviewed, debriefed, or queried through eliciting techniques. If they are interrogated, the reason is that they are known or believed to fall into one of the following categories.

2. Repatriates are sometimes interrogated, although other techniques are used more often. The proprietary interests of the host government will frequently dictate interrogation by a liaison service rather than by KUBARK. If KUBARK interrogates, the following preliminary steps are taken:
   a. A records check, including local and Headquarters traces.
   b. Testing of bona fides.
   c. Determination of repatriate's kind and level of access while outside his own country.
   d. Preliminary assessment of motivation (including political orientation), reliability, and capability as observer - and reporter.
   e. Determination of all intelligence or Communist relationships, whether with a service or party of the repatriate's own country, country of detention, or another. Full particulars are needed.

3. Defectors, escapees, and refugees are normally interrogated at sufficient length to permit at least a preliminary testing of bona fides. The experience of the post-war years has demonstrated that Soviet defectors (1) almost never defect solely or primarily because of inducement by a Western service, (2) usually leave the USSR for personal rather than ideological reasons, and (3) are often RIS agents. As a rule, Soviets seeking Western asylum are accorded the status of defectors because of their value as sources. they are customarily sent to a defector centér for detailed exploitation. Satellite escapees and refugees are handled as defectors only if they are highly knowledgeable and can satisfy established _intelligence

needs. All analyses of the defector-refugee flow have shown that the Orbit services are well-aware of the advantages offered by this channel as a means of planting their agents in target countries. Even the exodus of Hungarians on the heels of the 1956 uprising was exploited by the AVH. It is therefore important to remember that the bona fides of defectors cannot, as a rule, be established conclusively by interrogation alone. The cost in time and money precludes the intensive counterintelligence interrogation of all suspect defectors and refugees, but there is no sound alternative for selected cases.

4. Transferred sources referred to KUBARK by another service for interrogation are usually sufficiently well-known to the transferring service so that a file has been opened. Whenever possible, KUBARK should secure a copy of the file or its full informational equivalent before accepting custody.

5. Agents are more frequently debriefed than interrogated. If operational developments give rise to doubts about the security of a KUBARK agent or operation. It is recommended case officer use as an analytic tool. If it is then established or strongly suspected that the agent belongs to one of the following categories, further investigation and, eventually, interrogation usually follow.

a. Provocateur.

Many provocation agents are walk-ins posing as escapees, refugees, or defectors in order to penetrate emigre groups, ODYOKE intelligence, or other targets assigned by hostile services.

Although denunciations by genuine refugees and other evidence of information obtained from documents, local officials, and like sources may result in exposure, the detection of provocation frequently depends upon skilled interrogation. A later section of this manual deals with the preliminary testing of bona fides. But the results of preliminary testing are often inconclusive, and detailed interrogation is frequently essential to confession ~ and full revelation. Thereafter the provocateur may be questioned for operational and positive intelligence as well as counterintelligence provided that proper cognizance is taken of his status during the questioning and later, when reports are prepared.

b. Double agent.

The interrogation of DA's frequently follows a determination or strong suspicion that the double is "giving the edge" to the adversary service. As is also true for the interrogation of provocateurs, thorough preliminary investigation will pay handsome dividends when ; questioning gets under way. In fact, it is a basic principle of interrogation that the questioner should have at his disposal, before querying starts, as much pertinent information as can be gathered without the knowledge of the prospective interrogatee. KUBARK personnel who are planning interrogation of a suspect double agent may find it useful to consult

(b) penetration agent.

The goal of the penetration is to join a targeted group. Although the primary purpose of interrogation is the acquisition of information, a resistant source who has been "broken" should not be disregarded as a person when squeezed dry. All good interrogators avoid coercive techniques whenever the necessary information can be gained without them. In other words, physical or psychological duress is counter-productive when employed against a source whose voluntary cooperation can be enlisted without pressure. If coercion must be used and is successful, the temporary effect upon a hostile penetration agent, DA, or provocateur is the creation of a vacuum in his loyalties. He is likely to feel drained and apathetic. If the interrogator (or his service) restores the source's self-esteem at this point by supplying an acceptable rationalization for conversion to anti-Communist beliefs, the source will continue to volunteer cooperation. But if he has been compelled to divulge through the use of pressures exceeding his resistance (for example, narcosis or hypnosis), and if his motives are ignored once his information has been mined, he is likely to revert to the role of antagonist and try to cause us trouble by any means available to him. This topic is explored further in Part IX. of this manual. Swindlers and fabricators are usually interrogated for prophylactic reasons, not for counterintelligence information. The purpose is the prevention or nullification of damage to KUBARK, to other ODYOKE services, or to liaison. Swindlers and fabricators have little of CI significance to communicate but are notoriously skillful time wasters. Interrogation of them is usually inconclusive and, if prolonged, get unrewarding. The professional peddler with several IS contacts may prove an exception; but he will usually give the edge to a host security service because otherwise he cannot function with impunity.

B. Types of Sources: Personality Categories

The number of systems devised for categorizing human beings is large, and most of them are of dubious validity. Various categorical schemes are outlined in treatises on interrogation. The two typologies most frequently advocated are psychologic-emotional and geographic-cultural. Those who urge the former argue that the basic emotional-psychological patterns do not vary significantly with time, place, or culture. The latter school maintains the existence of a national character and sub-national categories, and interrogation guides based on this principle recommend approaches tailored to geographical cultures. It is plainly true that the interrogation source cannot be understood in a vacuum, isolated from social context. It is equally true that some of the most glaring blunders in interrogation (and other operational processes) have resulted from ignoring the source's background. Moreover, emotional-psychological schematizations sometimes present atypical extremes rather than the kinds of people commonly encountered by interrogators. Such typologies also cause disagreement even among professional psychiatrists and psychologists. Interrogators who adopt them and who note in an interrogatee one or two of the characteristics of "Type A" may mistakenly assign the source to Category A and assume the remaining traits. On the other hand, there are valid objections to the adoption of cultural-geographic categories for interrogation purposes (however valid they may be as KUCAGE concepts). The pitfalls of ignorance of the distinctive culture of the source have "received so much attention in recent years as to obscure somewhat the other tendency to think of persons from other cultures as more different from oneself than they actually are.

The interrogator is safest when he can proceed on the basis of an assumption that all individuals will react in essentially the same way to the same secret influence he employs....The populations of most nations are coming to share more of the outlook of their contemporaries in other nations than of their own national progenitors. Further, each large industrialized state produces occupational and social classes common to all such states. '"(3) The ideal solution would be to avoid all categorizing. Basically, all schemes for labelling people are wrong per se; applied arbitrarily, they always produce distortions. Every interrogator knows that a real understanding of the individual is worth far more than a thorough knowledge of this or that pigeon-hole to which he has been consigned. And for interrogation purposes the ways in 'which he differs from the abstract type may be more significant than the ways in which he conforms. But KUBARK does not dispose of the time or personnel to probe the depths of each source's individuality. In the opening phases of interrogation, or in a quick interrogation, we are compelled to make some use of the shorthand of categorizing, despite distortions. Like other interrogation aides, a scheme of categories is useful only if recognized for what it is--a set of labels that facilitate communication but are not the same as the persons thus labelled.

If an interrogatee lies persistently, an interrogator may report and dismiss him as a "pathological liar." Yet such persons may possess counterintelligence (or other) information quite equal in value to that held by other sources, and the interrogator likeliest to get at it is the man who is not content with labelling but is as interested in why the subject lies as in what he lies about. . With all of these reservations, then, and with the further observation that those who find these psychological-emotional categories pragmatically valuable should use them and those who do not should let them alone, the following nine types are described. The categories are based upon the fact that a person's past is always reflected, however dimly, in his present ethics and behavior. Old dogs can learn new tricks but not new ways of learning them. People do change, but what appears to be new behavior or a new psychological pattern is usually just a variant on the old theme.

It is not claimed that the classification system presented here is complete; some interrogatees will not fit into any one of the groupings. And like all other typologies, the system is plagued by overlap, so that some interrogatees will show characteristics of more than one group. Above all, the interrogator must remember that finding some of the characteristics of the group in a single source does not warrant an immediate conclusion that the source "belongs to'" the group, and that even correct labelling is not the equivalent of understanding people but merely an aid to understanding. The nine major groups within the psychological-emotional category adopted for this handbook are the following.

1. The orderly-obstinate character. People in this category are characteristically frugal, orderly, and cold; frequently they are quite intellectual. They are not impulsive in behavior. They tend to think things through logically and to act deliberately.

They often reach decisions very slowly. They are far less likely to make real are personal sacrifices for a cause than to use them as a temporary means of obtaining a permanent personal gain.

They are secretive and disinclined to confide in anyone else their plans and plots, which frequently concern the overthrow of some form of authority. They are also stubborn, although they may pretend cooperation or even believe that they are cooperating. They nurse grudges. The orderly -obstinate character considers himself superior to other people. Sometimes his sense of superiority is interwoven with a kind of magical thinking that includes all sorts of superstitions and fantasies about controlling his environment. He may even have a system of morality that is all his own. He sometimes gratifies his feeling of secret superiority by provoking unjust treatment. He also tries, characteristically, to keep open a line of escape by avoiding a any real commitment to anything. He is--and always has been--intensely concerned about his personal possessions. He is usually a tightwad who saves everything, has a strong sense of propriety, and is punctual and tidy. His money and other possessions have for him 3 a personalized quality; they are parts of himself He often carries around shiny coins, keepsakes, a bunch of keys, and other objects having for himself an actual or symbolic value.

Usually the orderly-obstinate character has a history of active rebellion in childhood, of persistently doing the exact opposite of what he is told to do. As an adult he may have learned to cloak his resistance and become passive-aggressive, but his determination to get his own way is unaltered. He has merely learned how to proceed indirectly if necessary. The profound fear and hatred of authority, persisting since childhood, is often well concealed in adulthood, For example, such a person may confess easily and quickly under interrogation, even to acts that he did not commit, in order to throw the interrogator off the trail of a significant discovery (or, more rarely, because of feelings of guilt). The interrogator who is dealing with an orderly-obstinate character should avoid the role of hostile authority. Threats and threatening gestures, table-pounding, pouncing on evasions or lies, and any similarly authoritative tactics will only awaken in such a subject his old anxieties and habitual defense mechanisms. To attain rapport, the interrogator should be friendly. It will probably prove rewarding if the room and the interrogator look exceptionally neat. Orderly-obstinate interroga-tees often collect coins or other objects as a hobby; time spent in sharing their interests may thaw some of the ice. Establishing rap-port is extremely important when dealing with this type. "Those personalities characterized by low originality, authoritarian ten-dencies, low achievement motivation, conventionality, and social dependence are among the apes estimated as being susceptible to manipulation in interrogation."

2. The optimistic character. This kind of source is almost con-stantly happy-go-lucky, impulsive, inconsistent, and undependable. He seems to enjoy a continuing state of well-being. He may be generous to a fault, giving to others as he wants to be given to. He may become an alcoholic or drug addict. He is not able to withstand very much pressure; he reacts to a challenge not by increasing his

efforts but rather by running away to avoid conflict. His convictions that "something will turn up", that "everything will work out all right", is based on his need to avoid his own responsibility for events and depend upon a kindly fate. , Such a person has usually had a great deal of over-indulgence in early life. He is sometimes the youngest member of a large family, the child of a middle-aged woman (a so-called "change -of-life baby"). If he has met severe frustrations in later childhood, he may be petulant, vengeful, and constantly demanding. As interrogation sources, optimistic characters respond best to a kindly, parental approach. If withholding, they can often be handled effectively by the Mutt-and-Jeff technique discussed later in this paper. Pressure tactics or hostility will make them retreat inside themselves, whereas reassurance will bring them out. They tend to seek promises, to cast the interrogator in the role of protector and problem-solver; and it is important that the interrogator avoid making any specific promises that cannot be fulfilled, because the optimist turned vengeful is likely to prove troublesome.

3. The greedy, demanding character. This kind of person affixes himself to others like a leech and clings obsessively. Although extremely dependent and passive, he constantly demands that others take care of him and gratify his wishes. If he considers himself wronged, he does not seek redress through his own efforts but tries to persuade another to take up the cudgels in his behalf--"'let's you and him fight.'" His loyalties are likely to shift whenever he feels that the sponsor whom he has chosen has let him down. Defectors of this type feel aggrieved because their desires were not satisfied in their countries of origin, but they soon feel equally deprived in a second land and turn against its government or representatives in the same way. The greedy and demanding character is subject to rather frequent depressions. He may direct a desire for revenge inward, upon himself; in extreme cases suicide may result. The greedy, demanding character often suffered from very early deprivation of affection or security. As an adult he continues to seek substitute parents who will care for him as his own, he feels, did a not. The interrogator dealing with a greedy, demanding character must be careful not to rebuff him; otherwise rapport will be destroyed. 4 On the other hand, the interrogator must not accede to demands which cannot or should not be met. Adopting the tone of an understanding father or big brother is likely to make the subject responsive. If he makes exorbitant requests, an unimportant favor may provide a satisfactory substitute because the demand arises not from a specific need but as an expression of the subject's need for

security. He is likely to find reassuring any manifestation of concern for his wellbeing. In dealing with this type--and to a considerable extent in dealing with any of the types herein listed--the interrogator must be aware of the limits and pitfalls of rational persuasion. If he seeks to induce cooperation by an appeal to logic, he should first determine whether the source's resistance is based on logic. The appeal will glance off ineffectually if the resistance is totally or chiefly emotional rather than rational. Emotional resistance can be dissipated only by emotional manipulation.

4. The anxious, self-centered character. Although this person is fearful, he is engaged in a constant struggle to conceal his fears. He is frequently a daredevil who compensates for his anxiety by pretending that there is no such thing as danger. He may be a stunt flier or circus performer who "proves" himself before crowds. He may also be a Don Juan. He tends to brag and often lies through hunger for approval or praise. Asa soldier or officer he may have been decorated for bravery; but if so, his comrades may suspect that his exploits resulted from a pleasure in exposing himself to danger and the anticipated delights of rewards, approval, and applause. The anxious, self-centered character is usually intensely vain and equally sensitive. People who show these characteristics are actually unusually fearful. The causes of intense concealed anxiety are too complex and subtle to permit discussion of the subject in this paper. Of greater importance to the interrogator than the causes is the opportunity provided by concealed anxiety for successful manipulation of the source. His desire to impress will usually be quickly evident. He is likely to be voluble. Ignoring or ridiculing his bragging, or cutting him short with a demand that he get down to cases, is likely to make him resentful and to stop the flow. Playing upon his vanity, especially by praising his courage, will usually be a successful tactic if employed skillfully. Anxious, self-centered interrogatees who are withholding significant facts, such as contact with a hostile service, are likelier to divulge if made to feel that the truth will not be used to harm them and if the interrogator also stresses the callousness and stupidity of the adversary in sending so valiant a person upon so ill-prepared a mission. There is little to be gained and much to be lost by exposing the non relevant lies of this kind of source. Gross lies about deeds of daring, sexual prowess, or other "proofs" of courage and manliness are best met with silence or with friendly but noncommittal replies unless they consume an inordinate amount of time. If operational use is contemplated, recruitment may sometimes be effected through such queries as, "I wonder if you would be willing to undertake a dangerous mission, "'

5. The guilt-ridden character. This kind of person has a strong cruel, unrealistic conscience. His whole life seems devoted to reliving his feelings of guilt. Sometimes he seems determined to atone; at other times he insists that whatever went wrong is the fault of somebody else. In either event he seeks constantly some proof or external indication that the guilt of others is greater than his own. He is often caught up completely in efforts to prove that he has been treated unjustly. In fact, he may provoke unjust treatment in order to assuage his conscience through punishment. Compulsive gamblers who find no real pleasure in winning but do find relief in losing belong to this class. So do persons who falsely confess to crimes. Sometimes such people actually commit crimes in order to confess and be punished. Masochists also belong in this category.

The causes of most guilt complexes are real or fancied wrongs done to parents or others whom the subject felt he ought to love and honor. As children such people may have been frequently scolded or punished, Or they may have been "model" children who repressed all natural hostilities. The guilt-ridden character is hard to interrogate. He may "confess" to hostile clandestine activity, or other acts of interest to KUBARK, in which he was not involved. Accusations levelled at him by the interrogator are likely to trigger such false confessions. Or he may remain silent when accused, enjoying the "punishment." He is a poor subject for LCFLUTTER. The complexities of dealing with conscience-ridden interrogatees vary so widely from case to case that it is almost impossible to list sound general principles. Perhaps the best advice is that the interrogator, once alerted by information from the screening process (see Part VI) or by the subject's excessive preoccupation with moral judgements, should treat as suspect and subjective any information provided by the interrogatee about any matter that is of moral concern to him. Persons with intense guilt feelings may cease resistance and cooperate if punished in some way, because of the gratification induced by punishment.

6. The character wrecked by success is closely related t to the guilt-ridden character. This sort of person cannot tolerate : success and goes through life failing at critical points. He is often accident-prone. Typically he has a long history of being promising and of almost completing a significant assignment or achievement but not bringing it off. The character who cannot stand success enjoys his ambitions as long as they remain fantasies but somehow ensures that they will not be fulfilled in reality. Acquaintances often feel that his success is just around : the corner, but something always

intervenes. In actuality this Pees something is a sense of guilt, of the kind described above. The person who avoids success has a conscience which forbids the pleasures of accomplishment and recognition. He frequently projects his guilt feelings and feels that all of his failures were someone else's fault. He may have a strong need to suffer and may seek danger or injury. As interrogatees these people who "cannot stand prosperity" pose no special problem unless the interrogation impinges upon their feelings of guilt or the reasons for their past failures. Then subjective distortions, not facts, will result. The successful interrogator will isolate this area of unreliability.

7. The schizoid or strange character lives in a world of fantasy much of the time. Sometimes he seems unable to distinguish reality from the realm of his own creating. The real world seems to him empty and meaningless, in contrast with the mysteriously significant world that he has made. He is extremely intolerant of any frustration that occurs in the outer world and deals with it by withdrawal into the interior realm. He has no real attachments to others, although he may attach symbolic and private meanings or values to other people. Children reared in homes lacking in ordinary affection and attention or in orphanages or state-run communes may become adults who belong to this category. Rebuffed in early efforts to attach themselves to another, they become distrustful of attachments and turn inward. Any link to a group or country will be undependable and, as a rule, transitory. At the same time the schizoid character needs external approval. Though he retreats from reality, he does not want to feel abandoned. As an interrogatee the schizoid character is likely to lie readily to win approval.

He will tell the interrogator what he thinks the interrogator wants to hear in order to win the award of seeing a smile on the interrogator's face. Because he is not always capable of distinguishing between fact and fantasy, he may be unaware of lying. The desire for approval provides the interrogator with a handle. Whereas accusations of lying or other indications of disesteem will provoke withdrawal from the situation teasing the truth out of the schizoid subject may not prove difficult if he is convinced that he will not incur favor through misstatements or disfavor through telling the truth. Like the guilt-ridden character, the schizoid character may be an unreliable subject for testing by LCFLUTTER because his internal needs lead him to confuse fact with fancy. He is also likely to make an unreliable agent because of his incapacity to deal with facts and to form real relationships. 8. The exception believes that the world owes him a great deal. He feels that he suffered a gross injustice, usually early in life, and should be repaid. Sometimes the injustice was meted out impersonally, by fate, as a physical deformity, an extremely painful illness or operation in childhood, or the early loss of one parent or both. Feeling that these misfortunes were undeserved, ' the exceptions regard them as injustices that someone or something must rectify. Therefore they claim as their right privileges not permitted others. When the claim is ignored or denied, the exceptions become rebellious, as adolescents often do. They are convinced that the justice of the claim is plain for all to see and that any refusal to grant it is willfully malignant. When interrogated, the exceptions are likely to make demands for money, resettlement aid, and other favors--demands that are completely out of proportion to the value of their contributions. Any ambiguous replies to such demands will be interpreted as acquiescence. Of all the types considered here, the exception is likeliest to carry an alleged injustice dealt him by KUBARK to the newspapers or the courts. The best general line to follow in handling those who believe that they are exceptions is to listen attentively (within reasonable time limits) to their grievances and to make no commitments that cannot be discharged fully.

Defectors from hostile intelligence services, doubles, provocateurs, and others who have had more than passing contact with a Sino-Soviet service may, if they belong to this category, prove unusually responsive to suggestions from the interrogator that they have been treated unfairly by the other service. Any planned operational are use of such persons should take into account the fact that they have ne no sense of loyalty to a common cause and are likely to turn aggrievedly against superiors. 9. The average or normal character is not a person wholly lacking in the characteristics of the other types. He may, in fact, exhibit most or all of them from time to time. But no one of them is persistently dominant; the average man's qualities of obstinacy, unrealistic optimism, anxiety, and the rest are not overriding or imperious except for relatively short intervals. Moreover, his reactions to the world around him are more dependent upon events in that world and less the product of rigid, subjective patterns than is true of the other types discussed.

C. Other Clues

Discusses in some detail the psychological characteristics of willing and unwilling DA's. This information will be useful to anyone who is about to interrogate a double agent.

The true defector (as distinguished from the hostile agent in defector's guise) is likely to have a history of opposition to authority. The sad fact is that defectors who left their homelands because they could not get along with their immediate or ultimate superiors are also likely to rebel against authorities in the new environment (a fact which usually plays an important part in re~defection). Therefore defectors are likely to be found in the ranks of the orderly-obstinate, the greedy and demanding, the schizoids, and the exceptions. Experiments and statistical analyses performed at the University of Minnesota concerned the relationships among anxiety and affiliative tendencies (desire to be with other people), on the one hand, and the ordinal position (rank in birth sequence) on the other. Some of the findings, though necessarily tentative and speculative, have some relevance to interrogation. (30). As is noted in the bibliography, the investigators concluded that isolation typically creates anxiety, that anxiety intensifies the desire to be with others who share the same fear, and that only and first-born children are more anxious and less willing or able to withstand pain than later-born children, Other applicable hypotheses are that fear increases the affiliative needs of first-born and only children much more than those of the later-born. These differences are more pronounced in persons from small families than in those who grew up in large families. Finally, only children are much likelier to hold themselves together and persist in anxiety producing situations than are the first-born, who more frequently try to retreat. In the other major respects - intensity of anxiety and emotional need to affiliate - no significant differences between "firsts" and "onlies"" were discovered. It follows that determining the subject's "ordinal position" before questioning begins maybe useful to the interrogator. But two cautions are in order, The first is that the findings are, at this stage, only tentative hypotheses, The second is that even if they prove accurate for large groups, the data are like those in actuarial tables; they have no specific predictive value for individuals.

VL SCREENING AND OTHER PRELIMINARIES

A. Screening

Defector reception centers and some large stations are able to conduct preliminary psychological screening before interrogation starts. The purpose of screening is to provide the interrogator, in advance, with a reading on the type and characteristics of the interrogatee. It is recommended that screening be conducted whenever personnel and facilities permit, unless it ¢ is reasonably certain that the interrogation will be of minor importance or that the interrogatee is fully cooperative. Screening should be conducted by interviewers, not interrogators; or at least the subjects should not be screened by the same KUBARK personnel who will interrogate them later. Other psychological testing aids are best administered by a trained psychologist.

Tests conducted on American POW's returned to U.S, jurisdiction in Korea during the Big and Little Switch suggest that prospective interrogatees who show normal emotional responsiveness on the Rorschach and related tests are likelier to prove cooperative under interrogation than are those whose responses indicate that they are apathetic and emotionally withdrawn or barren. Extreme resisters, however, share the response characteristics of collaborators; they differ in the nature and intensity of motivation rather than emotions. "An analysis of objective test records and biographical information is a sample of 759 Big Switch repatriates revealed that men who had collaborated differed from men who had not in the following ways: the collaborators were older, had completed more years of school, scored higher on intelligence tests administered after repatriation, had served longer in the Army prior to capture, and scored higher on the Psychopathic Deviate Scale - pd.... However, the 5 percent of the non collaborator sample who resisted actively - who were either decorated by the Army or considered to be 'reactionaries' by the Chinese - differed from the remaining group in precisely the same direction as the collaborator group and could not be distinguished from this group on any variable except age; the resisters were older than the collaborators." (33) Even a rough preliminary estimate, if valid, can be a boon to the interrogator because it will permit him to start with generally sound tactics from the beginning - tactics adapted to the personality of the source. Dr. Moloney has expressed the opinion, which we may use as an example of this, that the AVH was able to get what it wanted from Cardinal Mindszenty because the Hungarian service adapted its interrogation methods to his personality. "There can be no doubt that Mindszenty's preoccupation with the concept of becoming secure and powerful through the surrender of self to the greatest power of them all - his God idea - predisposed him to the response elicited in his experience with the communist intelligence. For him the surrender of self-system to authoritarian-system was natural, as was the very principle of martyrdom." (28) : The task of screening is made easier by the fact that the screener is interested in the subject, not in the information which he may possess. Most people--even many provocation agents who have been trained to recite a legend--will speak with some freedom ' about childhood events and familial relationships. And even the provocateur who substitutes a fictitious person for his real father will disclose some of his feelings about his father in the course of detailing his story about the imaginary substitute. If the screener has learned to put the potential source at ease, to feel his way along in each case, the source is unlikely to consider that a casual conversation about himself if dangerous.

The screener is interested in getting the subject to talk about himself. Once the flow starts, the screener should try not to stop it by questions, gestures, or other interruptions until sufficient information has been revealed to permit a rough determination of type. The subject is likeliest to talk freely if the screener's manner is friendly and patient. His facial expression should not reveal special interest in any one statement; he should just seem sympathetic and understanding. .Within a short time most people who have begun talking about themselves go back to early experiences, so that merely by listening and occasionally making a quiet, encouraging remark the screener can learn a great deal. Routine questions about school teachers, employers, and group leaders, for example, will lead the subject to reveal a good deal of how he feels about his parents, superiors, and others of emotional consequence to him because of associative links in his mind. It is very helpful if the screener can imaginatively place himself in the subject's position. The more the screener knows about the subject's native area and cultural background, the less likely is he to disturb the subject by an incongruous remark. Such comments as, "That must have been a bad time for you and your family, " or "Yes, I can see why you were angry," or "It sounds exciting" are sufficiently innocuous not to distract the subject, yet provide adequate evidence of sympathetic interest. Taking the subject's side against his enemies serves the same purpose, and such comments as "That was unfair; they had no right to treat you that way" will aid rapport and stimulate further revelations. It is important that gross abnormalities be spotted during the screening process, Persons suffering from severe mental illness will show major distortions, delusions, or hallucinations and will usually give bizarre explanations for their behavior. Dismissal or prompt referral of the mentally ill to professional specialists will save time and money. The second and related purpose of screening is to permit an educated guess about the source's probable attitude toward the sachet interrogation.

An estimate of whether the interrogatee will be cooperative or recalcitrant is essential to planning because very different methods are used in dealing with these two types. At stations or bases which cannot conduct screening in the formal sense, it is still worth-while to preface any important interrogation with an interview of the source, conducted by someone other than the interrogator and designed to provide a maximum of evaluative information before interrogation commences. Unless a shock effect is desired, the transition from the screening interview to the interrogation situation should not be abrupt. At the first meeting with the interrogatee it is usually a good idea for the interrogator to spend some time in the same kind of quiet, friendly exchange that characterized the screening interview. Even though the interrogator now has the screening product, the rough classification by type, he needs to understand the subject in his own terms. If he is immediately aggressive, he imposes upon the first interrogation session (and to a diminishing extent upon succeeding sessions) too arbitrary a pattern. As one expert has said, 'Anyone who proceeds without consideration for the disjunctive power of anxiety in human relationships will never learn interviewing."

B. Other Preliminary Procedures

The preliminary handling of other types of interrogation sources is usually less difficult. It suffices for the present purpose to list the following principles: .

1. All available pertinent information ought to be assembled and studied before the interrogation itself is planned, much less conducted. An ounce of investigation may be worth a pound of questions.

2. A distinction should be drawn as soon as possible between sources who will be sent to a defector reception center or) another site organized and equipped for interrogation and those whose interrogation will-be completed by the base or station with which contact is first established.

3. The suggested procedure for arriving at a preliminary assessment of walk-ins remains the same whether the walk-in is to be sent to a defector reception center or not. If the source is to be transferred to a center, it is helpful if the preliminary assessment of bona fides reaches the center before he does. The preliminary testing of bona fides by the station or base which; first takes up contact with a walk-in is discussed. The key points are repeated here for ease of reference. These preliminary tests are designed to supplement the technical examination of a walk-in's documents, substantive questions about claimed homeland or occupation, and other standard inquiries. The following questions, if asked, should be posed as soon as possible after the initial contact, while the walk-in is still under stress and before he has adjusted to a routine.

a. The walk-in may be asked to identify all : relatives and friends in the area, or even the country, in which PBPRIME asylum is first requested.

Traces should be run speedily. Provocation agents are : sometimes directed to "defect" in their target areas, " and friends or relatives already in place may be hostile assets.

b. At the first interview the questioner should be on the alert for phrases or concepts characteristic of intelligence or CP activity and should record such leads whether it is planned to follow them by interrogation on the spot or to refer them to an interrogation center for later exploitation.

c. LCFLUTTER should be used if feasible. If not, the walk-in may be asked to undergo such testing at a later date. Refusals should be recorded, as well as indications that the walk-in has been briefed on the technique by another service. The manner as well as the nature of the walk-in's reaction to the proposal should be noted.

If LCFLUTTER, screening, investigation, or any other methods do establish a prior intelligence history, the following minimal information should be obtained:

5. All documents that have a bearing on the planned interrogation merit study. Documents from Bloc countries, or those which are in any respect unusual or unfamiliar, are customarily sent to the proper field or headquarters component for technical analysis. 6. If during screening or any other pre-interrogation phase it is ascertained that the source has been interrogated before, this fact should be made known to the interrogator. Agents, for example, are accustomed to being questioned repeatedly and professionally. So are persons who have been arrested several times. People who

have had practical training in being interrogated become sophisticated subjects, able to spot uncertainty, obvious tricks, and other weaknesses.

C. Summary

Screening and the other preliminary procedures will help the interrogator - and his base, station, or center - to decide whether the prospective source (1) is likely to possess useful counterintelligence because of association with a foreign service or Communist Party and (2) is likely to cooperate voluntarily or not. Armed with these estimates and with whatever insights screening has provided into the personality of the source, the interrogator is ready to plan.

VII. PLANNING THE COUNTERINTELLIGENCE INTER-ROGATION

A. The Nature of Counterintelligence Interrogation

The long-range purpose of CI interrogation is to get from the source all the useful counterintelligence information that he has. The short-range purpose is to enlist his cooperation toward this end or, if he is resistant, to destroy his capacity for resistance and replace it with a cooperative attitude. The techniques used in nullifying resistance, inducing compliance, and eventually eliciting voluntary cooperation are discussed in Part VIII of this handbook. No two interrogations are the same. Every interrogation is shaped definitively by the personality of the source - and of the interrogator, because interrogation is an intensely interpersonal process. The whole purpose of screening and a major purpose of the first stage of the interrogation is to probe the strengths and weaknesses of the subject. Only when these have been established and understood does it become possible to plan realistically. Planning the CI interrogation of a resistant source requires an understanding (whether formalized or not) of the dynamics of confession. Here Horowitz's study of the nature of confession is pertinent. He starts by asking why confessions occur at all. "Why not always brazen it out when confronted by accusation? Why does a person convict himself through a confession, when, at the very worst, no confession would leave him at least as well off (and possibly better off)...?" He answers that confessions obtained without duress are usually the product of the following conditions:

1. The person is accused explicitly or implicitly and feels accused.

2. Asa result his psychological freedom - the extent to which he feels able to do what he wants to - is curtailed. This feeling need not correspond to confinement or any other external reality.

3. The accused feels defensive because he is on unsure ground. He does not know how much the accuser knows. Asa result the accused "has no formula for proper behavior, no role if you will, that he can utilize in this situation."

4. He perceives the accuser as representing authority. Unless he

believes that the accuser's powers far exceed his own, he is unlikely to feel hemmed in and defensive. And if he "perceives that the accusation is backed by 'real' evidence, the ratio of external forces to his own forces is increased and the person's psychological position is now more precarious. It is interesting to note that in such situations the accused tends toward over response, or exaggerated response; to hostility and emotional display; to self-righteousness, to counter accusation, to defense...."

5. He must believe that he is cut off from friendly or supporting forces. If he does, he himself becomes the only source of his "salvation."

6. "Another condition, which is most probably necessary, though not sufficient for confession, is that the accused person feels guilt. A possible reason is that a sense of guilt promotes self-hostility." It should be equally clear that if the person does not feel guilt he is not in his own mind guilty and will not confess to an act which others may regard as evil or wrong and he, in fact, considers correct. Confession in such a case can come only with duress even where all other conditions previously: mentioned may prevail."

7. The accused, finally, is pushed far enough along the path toward confession that it is easier for him to keep going than to turn back. He perceives confession as the only way out of his predicament and into freedom. (15) Horowitz has been quoted and summarized at some length because it is considered that the foregoing is a basically sound account of the processes that evoke confessions from sources whose resistance is not strong at the outset, who have not previously-been confronted with detention and interrogation, and who have not been trained by an adversary intelligence or security service in resistance techniques. A fledgling or disaffected Communist or agent, for example, might be brought to confession and cooperation without the use of any external coercive forces other than the interrogation situation itself, through the above-described progression of subjective events. It is important to understand that interrogation, as both situation and process, does of itself exert significant external pressure upon the interrogatee as long as he is not permitted to accustom himself to it. Some psychologists trace this effect back to infantile relationships. Meerlo, for example, says that every verbal relationship repeats to some degree the pattern of early verbal relationships between child and parent. (27) An interrogatee, in particular, is likely to see the interrogator as a parent or parent-symbol, an object of suspicion

and resistance or of submissive acceptance. If the interrogator is unaware of this unconscious process, the result can be a confused battle of submerged attitudes, in which the spoken words are often merely a-cover for the unrelated struggle being waged at lower levels of both personalities. On the other hand, the interrogator who does understand these facts and who knows how to turn them to his advantage may not need to resort to any pressures greater than those that flow directly from the interrogation setting and function. , Obviously, many resistant subjects of counterintelligence interrogation cannot be brought to cooperation, or even to compliance, merely through pressures which they generate within themselves or through the unreinforced effect of the interrogation situation.

Manipulative techniques - still keyed to the individual but brought to bear upon him from outside i himself - then become necessary. It is a fundamental hypothesis of this handbook that these techniques, which can succeed even with highly resistant sources, are in essence methods of inducing regression of the personality to whatever earlier and weaker level is required for the dissolution of resistance and the inculcation of dependence. All of the techniques employed to break through an interrogation roadblock, the entire spectrum from simple isolation to hypnosis and narcosis, are essentially ways of speeding up the process of regression. As the interrogatee slips back from maturity toward a more infantile state, his learned or structured personality traits fall away in a reversed chronological order, so that the characteristics most recently acquired - which are also the characteristics drawn upon by the interrogatee in his own defense - are the first to go.

As Gill and Brenman have pointed out, regression is basically a loss of autonomy. (13) Another key to the successful interrogation of the resisting source is the provision of an acceptable rationalization for yielding. As regression proceeds, almost all resisters feel the growing internal stress that results from wanting simultaneously to conceal and to divulge. To escape the mounting tension, the source may grasp at any face-saving reason for compliance - any explanation which will placate both his own conscience and the possible wrath of former superiors and associates if he is returned to Communist control. It is the business of the interrogator to provide the right rationalization at the right time. Here too the importance of understanding the interrogatee is evident; the right rationalization must be an excuse or reason that is tailored to the source's personality. The interrogation process is a continuum, and everything that takes place in the continuum influences all subsequent events, The continuing process, being interpersonal, is not

Therefore it is wrong to open a counterintelligence interrogation experimentally, intending to abandon unfruitful approaches one by one until a sound method is discovered by chance. The failures of the interrogator, his painful retreats from blind alleys, bolster the confidence of the source and increase his ability to resist. While the interrogator is struggling to learn from the subject the facts that should have been established before interrogation started, the subject is learning more and more about the interrogator.

B. *The Interrogation Plan*

Planning for interrogation is more important than the specifics of the plan. Because no two interrogations are alike, the interrogation cannot realistically be planned from A to Z, in all its particulars, at the outset. But it can and must be planned from A to F or A to M.

The chances of failure in an unplanned CI interrogation are unacceptably high. Even worse, a "dash-on-~regardless" approach can ruin the prospects of success even if sound methods are used later. The intelligence category to which the subject belongs, though not determinant for planning purposes, is still of some significance. The plan for the interrogation of a traveller differs from that for other types because the time available for questioning is often brief.

The examination of his bona fides, accordingly, is often less searching. He is usually regarded as reasonably reliable if his identity and freedom from other intelligence associations have been established, if records checks do not produce derogatory information, if his account of his background is free of omissions or discrepancies suggesting significant withholding, if he does not attempt to elicit information about the questioner or his sponsor, and if he willingly provides detailed information which appears reliable or is established as such.

Defectors can usually be interrogated unilaterally, at least for a time. Pressure for participation will usually come not from a foreign service but from an ODYOKE intelligence component. The time available for unilateral! testing and exploitation should be calculated at the outset, with a fair regard for the rights and interests of other members of the intelligence community. The most significant single fact to be kept in mind when planning the interrogation of Soviet defectors is that a certain percentage of them have proven to be controlled _agents; estimates of this percentage have ranged-as high as (b)(1) during a period of several years after 1955.

KUBARK's lack of executive powers is especially significant if the interrogation of a suspect agent or of any other subject who is expected to resist is under consideration. As a general rule, it is difficult to succeed in the CI interrogation of a resistant source unless the interrogating service can control the subject and his environment for as long as proves necessary.

As was mentioned earlier, agents and staff members of hostile services are often briefed about KUBARK's lack of police powers. Such sources may demand immediate release if detained for unilateral questioning. If the demand is refused, a they may later bring suit for illegal detention. Transfer to an interrogation center should not be used as an automatic solution. The interrogation plan of a station or base should take into account the legal considerations, problems of housing and guarding subjects undergoing unilateral questioning, and the frustration that may be engendered by expending much time and skilled manpower upon a recalcitrant source. Otherwise the station or base may press too hard in trying to quick results, wilt under pressure, and release an interrogatee from whom clarification has not been obtained.

C. The Specifics

1. The Specific Purpose Before questioning starts, the interrogator has clearly in mind what he wants to learn, why he thinks the source has the information, how important it is, and how it can best be obtained. Any confusion here, or any questioning based on the premise that the purpose will take shape after the interrogation is under way, is almost certain to lead to aimlessness and final failure. If the specific goals cannot be discerned clearly, further investigation is needed before querying starts.

2, Resistance The kind and intensity of anticipated resistance is estimated. It is useful to recognize in advance whether the information desired would be threatening or damaging in any way to the interests of the interrogatee. If so, the interrogator should consider whether the same information, or confirmation of it, can be gained from another source. Questioning suspects immediately, on a flimsy factual basis, will usually cause waste of time, not save it. On the other hand, if the needed information is not sensitive from the subject's viewpoint, merely asking for it is usually preferable to trying to trick him into admissions and thus creating an unnecessary battle of wits. The preliminary psychological analysis of the subject makes it easier to decide whether he is likely to resist and, if so, whether his resistance will be the product of fear that his personal interests will be damaged or the result of the non-cooperative nature of orderly-obstinate and related types.

The choice of methods to be used in overcoming resistance is also determined by the characteristics of the interrogatee.

3. The Interrogation Setting

The room in which the interrogation is to be conducted should be free of distractions. The colors of walls, ceiling, rugs, and furniture should not be startling. Pictures should be missing or dull. Whether the furniture should include a desk depends not upon the interrogator's convenience but rather upon the subject's anticipated reaction to connotations of superiority and officialdom. A plain table may be preferable. An overstuffed chair for the use of the interrogatee is sometimes preferable to a straight-backed, wooden chair because if he is made to stand for a lengthy period or is otherwise deprived of physical comfort, the contrast is intensified and increased disorientation results. Some treatises on interrogation are emphatic about the value of arranging the lighting so that its source is behind the interrogator and glares directly at the subject. Here, too, a flat rule is unrealistic. The effect upon a cooperative source is inhibitory, and the effect upon a withholding source may be to make him more stubborn. Like all other details, this one depends upon the personality of the interrogatee.

Good planning will prevent interruptions. If the room is also used for purposes other than interrogation, a "Do Not Disturb" sign or its equivalent should hang on the door when questioning is under way. The effect of someone wandering in because he forgot his pen or wants to invite the interrogator to lunch can be devastating. For the same reason there should not be a telephone in the room; it is certain to ring at precisely the wrong moment. Moreover, it is a visible link to the outside; its presence makes a subject feel less cutoff, better able to resist. The interrogation room affords ideal conditions for photographing the interrogatee without his knowledge by concealing a camera behind a picture or elsewhere. If a new safe house is to be used as the interrogation site, it should be studied carefully to be sure that the total environment can be manipulated as desired. For example, the electric current should be known in advance, so that transformers or other modifying devices will be on hand if needed. Arrangements are usually made to record the interrogation, transmit it to another room, or do both. Most experienced interrogators do not like to take notes. Not being saddled with this chore leaves them free to concentrate on what sources say, how they say it, and what else they do while talking or listening. Another reason for avoiding note taking is that it distracts and sometimes worries the interrogatee. In the course of several sessions conducted without note-taking, the subject is likely to fall into the comfortable illusion that he is not talking for the record. Another advantage of the tape is that it can be played back later. Upon some subjects the shock of hearing their own voices unexpectedly is unnerving. The record also prevents later twistings or denials of admissions. Tapes can also be edited and spliced, with effective results, if the tampering can be kept hidden. For example, if two suspects are involved and if B is merely told that A has confessed their joint duplicity, he may believe that the statement is a lie and that the interrogators are just up to their old tricks again. But if he hears A's taped confession, or A's tape edited to make it sound like a confession, the result may be quite different. A recording is also a valuable training aid for interrogators, who by this means can study their mistakes and their most effective techniques. Exceptionally instructive interrogations, or selected portions thereof, can also be used in the training of others. If possible, audio equipment should also be used to transmit the proceedings to another room, used as a listening post. The main advantage of transmission is that it enables the person in charge of the interrogation to note crucial points and map further strategy, replacing one interrogator with another, timing a dramatic interruption correctly, etc. It is also helpful to install a small blinker bulb behind the subject or to arrange some other method of signalling the interrogator, without the source's knowledge, that the questioner should leave the room for consultation or that someone else is about to enter.

4. The Participants

Interrogatees are normally questioned separately. Separation permits the use of a number of techniques that would not be possible otherwise. It also intensifies in the source the feeling of being cut off from friendly aid. Confrontation of two or more suspects with each other in order to produce recriminations or admissions is especially dangerous if not preceded by separate interrogation sessions which have evoked compliance from one of the interrogatees, or at least significant admissions involving both. Techniques for the separate interrogations of linked sources are discussed in Part IX. The number of interrogators used for a single interrogation case varies from one man to a large team. The size of the team depends on several considerations, Z chiefly the importance of the case and the intensity of source resistance. Although most sessions consist of one interrogator and one interrogatee, some of the techniques described later call for the presence of two, three, or four interrogators. The 4 two-man team, in particular, is subject to unintended antipathies and conflicts not called for by assigned roles. Planning and subsequent conduct should eliminate such cross-currents before they develop, especially because the source will seek to turn them to his advantage. Team members who are not otherwise engaged can be employed to best advantage at the listening post. Inexperienced interrogators find that listening to the interrogation while it is in progress can be highly educational. Once questioning starts, the interrogator is called upon to function at two levels. He is trying to do two seemingly contradictory things at once: achieve rapport with the subject but remain an essentially detached observer. Or he may project himself to the resistant interrogatee as powerful and ominous (in order to eradicate resistance and create the necessary conditions for rapport) while remaining wholly uncommitted at the deeper level, noting the significance of the subject's reactions and the effectiveness of his own performance. Poor interrogators often confuse this bi-level functioning with role-playing, but there is a vital difference. The interrogator who merely pretends, in his surface performance, to feel a given emotion or to hold a given attitude toward the source is likely to be unconvincing; the source quickly senses the deception. Even children are very quick to feel this kind of pretense. To be persuasive, the sympathy or anger must be genuine; but to be useful, it must not interfere with the deeper level of precise, unaffected observation. Bi-level functioning is not difficult or even unusual; most people act at times as both performer and observer unless their emotions are so deeply involved in the situation that the critical faculty disintegrates.

Through experience the interrogator becomes adept in this dualism. The interrogator who finds that he has become emotionally involved and is no longer capable of unimpaired objectivity should report the facts so that a substitution can be made. Despite all . planning efforts to select an interrogator whose age, background, skills, personality, and experience make him the best choice for the job, it sometimes happens that both questioner and subject feel, when they first meet, an immediate attraction or antipathy which is so strong that a change of interrogators quickly becomes essential. No interrogator should be reluctant to notify his superior when emotional involvement becomes evident. Not the reaction but a failure to report it would be evidence of a lack of professionalism. Other reasons for changing interrogators should be anticipated and avoided at the outset. During the first part of the interrogation the developing relationship between the questioner and the initially uncooperative source is more important than the information obtained; when this relationship is destroyed by a change of interrogators, the replacement must start nearly from scratch. In fact, he starts with a handicap, because exposure to interrogation will have made the source a more effective resister. Therefore the base, station, or center should not assign as chief interrogator a person whose availability will end before the estimated completion of the case. 5. The Timing Before interrogation starts, the amount of time probably required and probably available to both interrogator and interrogatee should be calculated. If the subject is not to be under detention, his normal schedule is ascertained in advance, so that he will not have to be released at a critical point because he has an appointment or has to go to work. Because pulling information from a recalcitrant subject is the hard way of doing business, interrogation should not begin until all pertinent facts available from overt and from , cooperative sources have been assembled. Interrogation sessions with a resistant source who is under detention should not be held on an unvarying schedule. The capacity for resistance is diminished by disorientation. '

The subject may be left alone for days; and he may be returned to his cell, allowed to sleep for five minutes, and brought back to an interrogation which is conducted as though eight hours had intervened. The principle is that sessions should be so planned as to disrupt the source's sense of chronological order. 6. The Termination The end of an interrogation should be planned before questioning starts. The kinds of questions asked, the methods employed, and even the goals sought may be shaped by what | will happen when the end is reached. If, for example, the subject is to be turned over to a host service, it becomes more than usually important to hold to a minimum the amount of information about KUBARK and its methods that he can communicate. If he is to be released upon the local economy, perhaps blacklisted as a suspected hostile agent but not subjected to subsequent counterintelligence surveillance, it is important to avoid an inconclusive ending that has warned the interrogatee of our doubts but has established nothing. The poorest interrogations are those that trail off into an inconclusive nothingness. A number of practical terminal details should also be considered in advance. Are the source's documents to be returned to him, and will they be available in time? Is he to be paid? If he is a fabricator or hostile agent, has he been photographed and fingerprinted? Are subsequent contacts necessary or desirable, and have recontact provisions been arranged? Has a quit-claim been obtained? As was noted at the beginning of this section, the successful interrogation of a strongly resistant source ordinarily involves two key processes: the calculated regression of the interrogatee and the provision of an acceptable rationalization. If these two steps have been taken, it becomes very important to clinch the new tractability by means of conversion. In other words, a subject who has finally divulged the information sought and who has been given a reason for divulging which salves his self-esteem, his conscience, or both, will often be in a mood to take the final step of accepting the interrogator's values and making common cause with him. If operational use is now sachet contemplated, conversion is imperative. But even if the source has no further value after his fund of information has been mined, spending some extra time with him in order to replace his new ; sense of emptiness with new values can be good insurance, All ; non-Communist services are bothered at times by disgruntled ex-interrogatees who press demands and threaten or take hostile action if the demands are not satisfied. Defectors in particular, because they are often hostile toward any kind of authority, cause trouble by threatening or bringing suits in local courts, arranging publication of vengeful stories, or going to the local police, The former interrogatee is especially likely to be a future trouble-maker if during interrogation he was subjected to a form of compulsion imposed from outside himself. Time spent, after the interrogation ends, in fortifying the source's sense of acceptance in the interrogator's world may be only a fraction of the time required to bottle up his attempts to gain revenge. Moreover, conversion may create a useful and enduring asset. (See also remarks in VIII B 4.)

VIII. THE NON-COERCIVE COUNTERINTELLIGENCE INTERROGATION

A, General Remarks

The term non-coercive is used above to denote methods of inter-rogation that are not based upon the coercion of an unwilling subject through the employment of superior force originating outside himself. However, the non-coercive interrogation is not conducted without pressure. On the contrary, the goal is to gener-ate maximum pressure, or at least as much as is needed to induce compliance. The difference is that the pressure is generated inside the interrogatee.

His resistance is sapped, his urge to yield is fortified, until in the end he defeats himself. Manipulating the subject psychologically until he becomes compliant, without applying external methods of forcing him to submit, sounds harder than it is. The initial advantage lies with the interrogator. From the outset, he knows a great deal more about the source than the source knows about him. And he can create and amplify an effect of omniscience in a number of ways.

For example, he can show the interrogatee a thick file bearing his own name, Even if the file contains little or nothing but blank paper, the air of familiarity with which the interrogator refers to the subject's background can convince some sources that all is known and that resistance is futile. If the interrogatee is under detention, the interrogator can also manipulate his environment. Merely by cutting off all other human contacts, "the interrogator monopolizes the social environment of the source."(3) He exercises' the powers of an all-powerful parent, determining when the source will be sent to bed, when and what he will eat, whether he will be rewarded for good behavior or punished for being bad. The interrogator can and does make the subject's world not only unlike the world to which he had been accustomed but also strange in itself - a world in which familiar patterns of time, space, and sensory perception are overthrown. He can shift the environment abruptly. For example, a source who refuses to talk at all can be placed in unpleasant soli-tary confinement for a time. Then a friendly soul treats him to an unexpected walk in the woods. Experiencing relief and exhilara-tion, the subject ' will usually find it impossible not to respond to innocuous comments on the weather and the flowers, These are expanded to include reminiscences, and soon a precedent of verbal exchange has been established, Both the Germans and the Chinese have used this trick effectively.. The interrogator also chooses the emotional key or keys in which the interrogation or any part of it will be played. Because of these and other advantages, "... skilled and determined interrogators are almost invariably successful in eliciting some information from their sources.... For prisonerof-war interrogation, the figures generally given as the proportion of sources who abandon the 'name, rank, number only' rule, or other injunctions of silence, are between 95 and 100 percent... .

B. The Structure of the Interrogation

A counterintelligence interrogation consists of four parts: the opening, the reconnaissance, the detailed questioning and the conclusion.

1. The Opening

Most resistant interrogatees block off access to significant counterintelligence in their possession for one or more of : four reasons. The first is a specific negative reaction to the interrogator, Poor initial handling or a fundamental antipathy can make a source uncooperative even if he has nothing significant or damaging to conceal. The second cause is that some sources are resistant "by nature" - i.e. by early conditioning - to any compliance with authority. The third is that the subject believes that the information sought will be damaging or incriminating for him personally, that cooperation with the interrogator will have consequences more painful for him than the results of non-cooperation. The fourth is ideological resistance. The source has identified himself with a cause, a political movement or organization, or an opposition intelligence service. Regardless of his attitude toward the interrogator, his own personality, and his fears for the future, the person who is deeply devoted to a hostile cause will ordinarily prove strongly resistant under interrogation. A principal goal during the opening phase is to confirm the personality assessment obtained through screening and to allow the interrogator to gain a deeper understanding of the source as an individual, Unless time is crucial, the interrogator should not become impatient if the interrogatee wanders from the purposes of the interrogation and reverts to personal concerns, Significant facts not produced during screening may be revealed. The screening report itself is brought to life, the type becomes an individual, as the subject talks. And sometimes seemingly rambling monologues about personal matters are preludes to significant admissions, Some people cannot bring themselves to provide information that puts them in an unfavorable light until, through a lengthy prefatory rationalization, they feel that they have set the stage, that the interrogator will now understand why they acted as they did. If face-saving is necessary to the interrogatee, it will be a waste of time to try to force him to cut the preliminaries short and get down to cases, In his view, he is dealing with the important topic, the why. He will be offended and may become wholly uncooperative if faced with insistent demands for the naked what. There is another advantage in letting the subject talk freely and even ramblingly in the first stage of interrogation. The interrogator is free to observe. Human beings communicate a great deal by non-verbal means. Skilled interrogators, for example, listen closely to voices and learn a great deal from them. An interrogation is not merely a verbal performance; it is a vocal performance, and the voice projects tension, fear, a dislike of certain topics, and other useful pieces of information. It is also helpful to watch the subject's mouth, which is as a rule much more revealing than his eyes. Gestures and postures also tell a story. If a subject normally gesticulates broadly at times and is at other times physically relaxed but at some point sits stiffly motionless, his posture is likely to be the physical image of his mental tension, The interrogator should make a mental note of the topic that caused such a reaction. One textbook on interrogation lists the following physical indicators of emotions and recommends that interrogators note them, not as conclusive proofs but as assessment aids:

(1) A ruddy or flushed face is an indication of anger or embarrassment but not necessarily of guilt.

(2) A "cold sweat" is a strong sign of fear and shock.

(3) A pale face indicates fear and usually shows that the interrogator is hitting close to the mark.

(4) A dry mouth denotes nervousness.

(5) Nervous tension is also shown by wringing a handkerchief or clenching the hands tightly.

(6) Emotional strain or tension may cause a pumping of the heart which becomes visible in the pulse and throat.

(7) A alight gasp, ·holding the breath. or an unsteady voice may betray the subject.

(8) Fidgeting may take many forms, all of which are good indications of nervousness.

(9) A man under emotional strain or nervous tension will involuntarily draw his elbows to his sides. It is a protective defense mechanism.

(10) The movement of the foot when one leg is crossed over the

knee of the other can serve as an indicator, The circulation of the blood to the lower leg is partially cut off, thereby causing a slight lift or movement of the free foot with each heart beat. This becomes more pronounced and observable as the pulse rate increases.

Pauses are also significant. Whenever a person is talking about a subject of consequence to himself, he goes through a process of advance self-monitoring, performed at lightning speed. This self-monitoring is more intense if the person is talking to a stranger and especially intense if he is answering the stranger's questions. Its purpose is to keep from the questioner any guilty information or information that would be damaging to the speaker's self-esteem. When questions or answers get close to sensitive areas, the pre-scanning is likely to create mental blocks. These in turn produce unnatural pauses, meaningless sounds designed to give the speaker more time, or other interruptions, It is not easy to distinguish between innocent blocks -- things held back for reasons of personal prestige -- and guilty blocks -- things the interrogator needs to know. But the successful establishment of rapport will tend to eliminate innocent blocks, or at least to keep them to a minimum.

The establishment of rapport is the second principal purpose of the opening phase of the interrogation, Sometimes the interrogator knows in advance, as a result of screening, that the subject will be uncooperative. At other times the probability of resistance is established without screening; detected hostile agents, for example, usually have not only the will to resist but also the means, through a cover story or other explanation. But the anticipation of withholding increases rather than diminishes, the value of rapport. In other words, lack of rapport may cause an interrogatee to withhold information that he would otherwise provide freely, whereas the existence of rapport may induce an interrogatee who is initially determined to withhold to change his attitude. Therefore the interrogator must not become hostile if confronted with initial hostility, or in any other way confirm such negative attitudes as he may encounter at the outset. During this first phase his attitude should remain business-like but also quietly (not ostentatiously) friendly and welcoming. Such opening remarks by subjects as, "I know what you so-and-so's are after, and I can tell you right now that you're not going to get it from me" are best handled by an unperturbed "Why don't you tell me what has made you angry?" At this stage the interrogator should avoid being drawn into conflict, no matter how provocative may be the attitude or language of the interrogatee, If he meets truculence with neither insincere protestations that he is the subject's "pal" nor an equal anger but rather a calm interest in what has aroused the subject, the interrogator has gained two advantages right at the start. He has established the superiority that he will need later, as the questioning develops, and he has increased the chances of establishing rapport. How long the opening phase continues depends upon how long it takes to establish rapport or to determine that voluntary cooperation is unobtainable. It may be literally a matter of seconds, or it may be a drawn-out, up-hill battle. Even though the cost in time and patience is sometimes high, the effort to make the subject feel that his questioner is a sympathetic figure should not be abandoned until all reasonable resources have been exhausted (unless, of course, the interrogation does not merit much time), Otherwise, the chances are that the interrogation will not produce optimum results. In fact, it is likely to be a failure, and the interrogator should not be dissuaded from the effort to establish rapport by an inward conviction that no man in his right mind would incriminate himself by providing the kind of information that is sought. The history of interrogation is full of confessions and other self-incriminations that were in essence the result of a substitution of the interrogation world for the world outside, other words, as the sights and sounds of an outside world fade away, its significance for the interrogatee tends to do likewise. That world is replaced by the interrogation room, its two occupants, and the dynamic relationship between them. As interrogation goes on, the subject tends increasingly to divulge or withhold in accordance with the values of the interrogation world rather than those of the outside world (unless the periods of questioning are only brief interruptions in his normal life). In this small world of two inhabitants a clash of personalities -- as distinct from a conflict of purposes -assumes exaggerated force, like a tornado in a wind-tunnel.

The self-esteem of the interrogatee and of the interrogator becomes involved, and the interrogatee fights to keep his secrets from his opponent for subjective reasons, because he is grimly determined not to be the loser, the inferior. If on the other hand the interrogator establishes rapport, the subject may withhold because of other reasons, but his resistance often lacks the bitter, last-ditch intensity that results if the contest becomes personalized. The interrogator who senses or determines in the opening phase that what he is hearing is a legend should resist the first, natural impulse to demonstrate its falsity. In some interrogatees the ego-demands, the need to save face, are so intertwined with preservation of the cover story that calling the man a liar will merely intensify resistance, It is better to leave an avenue of escape, a loophole which permits the source to correct his story without looking foolish, If it is decided, much later in the interrogation, to confront the interrogatee with proof of lying, the following related advice about legal cross-examination may prove helpful. "Much depends upon the sequence in which one conducts the cross-examination of a dishonest witness. You should never hazard the important question until you have laid the foundation for it in such a way that, when confronted with the fact, the witness can neither deny nor explain it. One often sachet sees the most damaging documentary evidence, in the forms of letters or affidavits, fall absolutely flat as betrayers of falsehood, merely because of the unskillful way in which they are handled. If you have in your possession a letter written by the witness, in which he takes an opposite position on some part of the case to the one he has just sworn to, avoid the common error of showing the witness the letter for identification, and then reading it to him with the inquiry, 'What have you to say to that?' During the reading of his letter the witness will be collecting his thoughts and getting ready his explanations in anticipation of the question that is to follow, and the effect of the damaging letter will be lost.... The correct method of using such a letter is to lead the witness quietly into repeating the statements he has made in his direct testimony, and which his letter contradicts. Then read it off to him. The witness has [no explanation]. He has stated the fact, there is nothing to qualify.

2. The Reconnaissance

If the interrogatee is cooperative at the outset or if rapport is established during the opening phase and the source becomes cooperative, the reconnaissance stage is needless; the interrogator proceeds directly to detailed questioning. But if the interrogatee is withholding, a period of exploration is necessary. Assumptions have normally been made already as to what he is withholding: that he is a fabricator, or an RIS agent, or something else he deems it important to conceal, Or the assumption may be that he had knowledge of such activities carried out by someone else. At any rate, the purpose of the reconnaissance is to provide a quick testing of the assumption and, more importantly, to probe the causes, extent, and intensity of resistance. During the opening phase the interrogator will have charted the probable areas of resistance by noting those topics 4 which caused emotional or physical reactions, speech blocks, or other indicators. He now begins to probe these areas.

Every experienced interrogator has noted that if an interrogatee is withholding, his anxiety increases as the questioning nears the mark. The safer the topic, the more voluble the source. But as the questions make him increasingly uncomfortable, the interrogatee becomes less communicative or perhaps even hostile. During the opening phase the interrogator has gone along with this protective mechanism. Now, however, he keeps coming back to each area of sensitivity until he has determined the location of each and the intensity of the defenses. If resistance is slight, mere persistence may overcome it; and detailed questioning may follow immediately. But if resistance is strong, a new topic should be introduced, and detailed questioning reserved for the third stage. Two dangers are especially likely to appear during the reconnaissance. Up to this point the interrogator has not continued a line of questioning when resistance was encountered. Now, however, he does so, and rapport may be strained. Some interrogatees will take this change personally and tend to personalize the conflict. The interrogator should resist this tendency. If he succumbs to it, and becomes engaged in a battle of wits, he may not be able to accomplish the task at hand. The second temptation to avoid is the natural inclination to resort prematurely to ruses or coercive techniques in order to settle the matter then and there. The basic purpose of the reconnaissance is to determine the kind and degree of pressure that will be needed in the third stage. The interrogator should reserve his fire-power until he knows what he is up against.

3. The Detailed Questioning

a. If rapport is established and if the interrogatee has nothing significant to hide, detailed questioning presents only routine problems. The major routine considerations are the following: The interrogator must know exactly what he wants to know. He should have on paper or firmly in mind all the questions to which he seeks answers.

It usually happens that the source has a relatively large body of information that has little or no intelligence value and only a small collection of nuggets. He will naturally tend to talk about what he knows best. The interrogator should not show quick impatience, but neither should he allow the results to get out of focus. The determinant remains what we need, not what the interrogatee can most readily provide. At the -same time it is necessary to make every effort to keep the subject from learning through the interrogation process precisely where our informational gaps lie. This principle is especially important if the interrogatee is following his normal life, going home each evening and appearing only once or twice a week for questioning, or if his bona fides remains in doubt. Under almost all circumstances, however, a clear revelation of our interests and knowledge should be avoided. It is usually a poor practice to hand to even the most cooperative interrogatee an orderly list of questions and ask him to write the answers. (This stricture does not apply to the writing of autobiographies or on informational matters not a subject of controversy with the source.) Some time is normally spent on matters of little or no intelligence interest for purposes of concealment. The interrogator can abet the process by making occasional notes -- or pretending to do so -- on items that seem important to the interrogatee but are not of intelligence value. From this point of view an interrogation can be deemed successful if a source who is actually a hostile agent can report to the opposition only the general fields of our interest but cannot pinpoint specifics without including misleading information.

It is sound practice to write up each interrogation report on the day of questioning or, at least, before the next session, so that defects can be promptly remedied and gaps or contradictions noted in time. It is also a good expedient to have the interrogatee make notes of topics that should be covered, which occur to him while discussing the immediate matters at issue. The act of recording the stray item or thought on paper fixes it in the interrogatee's mind. Usually topics popping up in the course of an interrogation are forgotten if not noted; they tend to disrupt the interrogation plan if covered by way of digression on the spot. Debriefing questions should usually be couched to provoke a positive answer and should be specific. The questioner should not accept a blanket negative without probing. For example, the question "Do you know anything about Plant X?" is likelier to draw a negative answer then 'Do you have any friends who work at Plant X?" or "Can you describe its exterior?" It is important to determine whether the subject's knowledge of any topic was acquired at first hand, learned indirectly, or represents merely an assumption. If the information was obtained indirectly, the identities of sub-sources and related information about the channel are needed. If statements rest on assumptions, the facts upon which the conclusions are based are necessary to evaluation.

As detailed questioning proceeds, additional biographical data will be revealed. Such items should be entered into the record, but it is normally preferable not to diverge from an impersonal topic in order to follow a biographical lead. Such leads can be taken up later unless they raise new doubts about bona fides. As detailed interrogation continues, and especially at the half-way mark, the interrogator's desire to complete the task may cause him to be increasingly business-like or even brusque. He may tend to curtail or drop the usual inquiries about the subject's well-being with which he opened earlier sessions. He may feel like dealing more and more abruptly with reminiscences or digressions. His interest has shifted from the interrogatee himself, who just a while ago was an interesting person, to the task of getting at what he knows, But if rapport has been established, the interrogatee will be quick to sense and resent this change of attitude. This point is particularly important if the interrogatee is a defector faced with bewildering changes and in a highly emotional state. Any interrogatee has his ups and downs, times when he is tired or half-ill, times when his personal problems have left his nerves frayed. The peculiar intimacy of the interrogation situation and the very fact that the interrogator has deliberately fostered rapport will often lead the subject to talk about his doubts, fears, and other personal reactions. The interrogator should neither cut off this flow abruptly nor show impatience unless it takes up an inordinate amount of time or unless it seems likely that all the talking about personal matters is being used deliberately as a smoke screen to keep the interrogator from doing his job. If the interrogatee is believed cooperative, then from the beginning to the end of the process he should feel that the interrogator's interest in him has remained constant. Unless the interrogation is soon over, the interrogatee's attitude toward his questioner is not likely to remain constant. He will feel more and more drawn to the questioner or increasingly antagonistic. Asa rule, the best way for the interrogator to keep the relationship on an even keel is to maintain the same quiet, relaxed, and open-minded attitude from start to finish. Detailed interrogation ends only when (1) all useful counterintelligence information has been obtained; (2) diminishing returns and more pressing commitments compel a cessation; or (3) the base, station, or center admits full or partial defeat. Termination for any reason other than the first is only temporary. It is a profound mistake to write off a success-fully resistant interrogatee or one whose questioning was ended before his potential was exhausted. KUBARK must keep track of such persons, because people and circumstances change. Until the source dies or tells us everything that he knows that is pertinent to our purposes, his interrogation may be interrupted, perhaps for years -- but it has not been completed,

4. The Conclusion

The end of an interrogation is not the end of the interrogator's responsibilities. From the beginning of planning to the end of questioning it has been necessary to understand and guard against the various troubles that a vengeful ex-source can cause. As was pointed out earlier, KUBARK's lack of executive authority abroad and its operational need for facelessness make it peculiarly vulner-able to attack in the courts or the press. The best defense against such attacks is prevention, through enlistment or enforcement of compliance. However real cooperation is achieved, its existence seems to act as a deterrent to later hostility. The initially resistant subject may become cooperative because of a partial identification with the interrogator and his interests, or the source may make such an identification because of his cooperation. In either event, he is unlikely to cause serious trouble in the future. Real difficulties are more frequently created by interrogatees who have succeeded in withholding, The following steps are normally a routine part of the conclusion:

Personal property is returned to the interrogatee against receipt. If something cannot be returned at the time -- 2 document, for example -- an explanation or settlement satisfactory to the source is made if possible. If the source is to be rewarded by cash or a gift, a receipt is normally obtained, e. If during the final session the interrogatee manifests serious hostility, threatens court action, or otherwise indicates an intention to seek revenge, Headquarters is promptly notified. The interrogator participates in formulating the disposal plan, because of the relevance of his intimate knowledge of the source.

C. Techniques of Non-Coercive Interrogation of Resistant Sources

If source resistance is encountered during screening or during the opening or reconnaissance phases of the interrogation, non coercive methods of sapping opposition and strengthening the tendency to yield and to cooperate may be applied. Although these methods appear here in an approximate order of increasing pressure, it should not be inferred that each is to be tried until the key fits the lock. On the contrary, a large part of the skill and the success of the experienced interrogator lies in his ability to match method to source. The use of unsuccessful techniques will of itself increase the interrogatee's will and ability to resist. This principle also affects the decision to employ coercive techniques and governs the choice of these methods. If in the opinion of the interrogator a totally resistant source has the skill and determination to with- stand any non-coercive method or combination of methods, it is better to avoid them completely. The effectiveness of most of the non-coercive techniques depends . upon their unsettling effect. The interrogation situation is in itself disturbing to most people encountering it for the first time. The aim is to enhance this effect, to disrupt radically the familiar emotional and psychological asso- ciations of the subject. When this aim is achieved, resistance is seriously impaired. There is an interval -which may be extremely brief -- of suspended animation, a kind of psychological shock or paralysis, It is caused by a traumatic or sub-traumatic experience which explodes, as it were, the world that is familiar to the subject as well as his image of himself within that world. Experienced interrogators recognize this effect when it appears and know that at this moment the source is far more open to suggestion, far likelier to comply, than he was just before he experienced the shock, Another effect frequently produced by non-coercive (as well as coercive) methods is the evocation within the interrogatee of feelings of guilt. Most persons have areas of guilt in their emotional topographies, and an interrogator can often chart these areas just by noting refusals to follow certain lines of questioning. Whether the sense of guilt has real or imaginary causes does not affect the result of intensification of guilt feelings. Making a person feel more and more guilty normally increases both his anxiety and his urge to cooperate as a means of escape. In brief, the techniques that follow should match the personality of the individual interrogatee, and their effectiveness is intensified by good timing and rapid exploitation of the moment of shock. (A few of the following items are drawn from Sheehan.)

1. Going Next Door

Occasionally the information needed from a recalcitrant interrogatee is obtainable from a willing source. The interrogator should decide whether a confession is essential to his purpose or whether information which may be held by others as well as the unwilling source is really his goal. The labor of extracting the truth from unwilling interrogatees should be undertaken only if the same information is not more easily obtainable elsewhere or if operational considerations require self-incrimination.

2. Nobody Loves You

An interrogatee who is withholding items of no grave : consequence to himself may sometimes be persuaded to talk by the simple tactic of pointing out that to date all of the information about his case has come from persons other than himself. The interrogator wants to be fair. He recognizes that some of the denouncers may have been biased or malicious. In any case, there is bound to be some slanting of the facts unless the interrogatee redresses the balance. The source owes it to himself to be sure that the interrogator hears both sides of the story.

3. The All-Seeing Eye (or Confession is Good for the Soul)

The interrogator who already knows part of the story explains to the source that the purpose of the questioning is not to gain information; the interrogator knows everything already. His real purpose is to test the sincerity (reliability, honor, etc.) of the source, The interrogator then asks a few questions to which he knows the answers. If the subject lies, he is informed firmly and dispassionately that he has lied. By skilled manipulation of the known, the questioner can convince a naive subject that all his secrets are out and that further : resistance would be not only pointless but dangerous. If this | technique does not work very quickly, it must be dropped before the interrogatee learns the true limits of the questioner's knowledge.

4. The Informer

Detention makes a number of tricks possible. One of these, planting an informant as the source's cell mate, is so well-known, especially in Communist countries, that its usefulness is impaired if not destroyed. Less well known is the trick of planting two informants in the cell. One of them, A, tries now and then to pry a little information from the source; B remains quiet. At the proper time, and during A's absence, B warns the source not to tell A anything because B suspects him of being an informant planted by the authorities.

Suspicion against a single cell mate may sometimes be broken down if he shows the source a hidden microphone that he has "found" and suggests that they talk only in whispers at the other end of the room.

5. News from Home

Allowing an interrogatee to receive carefully selected letters from home can contribute to effects desired by the interrogator. Allowing the source to write letters, especially if he can be led to believe that they will be smuggled out without the knowledge of the authorities, may produce information which is difficult to extract by direct questioning.

6. The Witness

If others have accused the interrogatee of spying for a hostile service or of other activity which he denies, there is a temptation to confront the recalcitrant source with his accuser or accusers, But a quick confrontation has two weaknesses: it is likely to intensify the stubbornness of denials, and it spoils the chance to use more subtle methods. One of these is to place the interrogatee in an outer office and escort past him, and into the inner office, an accuser whom he knows personally or, in fact, any person -even one who is friendly to the source and uncooperative with the interrogators -- who is believed to know something about whatever the interrogatee is concealing.

It is also essential that the interrogatee know or suspect that the witness may be in possession of the incriminating information. The witness is whisked past the interrogatee; the two are not allowed to speak to each other. A guard and a stenographer remain in the outer office with the interrogatee. After about an hour the inter- rogator who has been questioning the interrogatee in past sessions opens the door and asks the stenographer to come in, with steno pad and pencils. After a time she re-emerges and types material from her pad, making several carbons. She pauses, points at the interrogatee, and asks the guard how his name is spelled. She may also ask the interrogatee directly for the proper spelling of a street, a prison, the name of a Communist intelligence officer, or any other factor closely linked to the activity of which he is accused. She takes her completed work into the inner office, comes back out, and telephones a request that someone come up to act as legal witness.

Another man appears and enters the inner office, The person cast in the informer's role may have been let out a back door at the beginning of these proceedings; or if cooperative, he may continue his role. In either event, a couple of interrogators, with or without the "informer", now emerge from the inner office. In contrast to their earlier demeanor, they are now relaxed and smiling. The interrogator in charge says to the guard, "O.K,, Tom, take hirn back, We don't need him any more." Even if the interrogatee now insists on telling his side of the story, he is told to relax, because the interrogator will get around to him tomorrow or the next day. A session with the witness may be recorded. If the witness denounces the interrogatee, there is no problem. If he does not, the interrogator makes an effort to draw him out about a hostile agent recently convicted in court or otherwise known to the witness, During the next interrogation session with the source, a part of the taped denunciation can be played back to him if necessary. Or the witnesses' remarks about the known spy, edited as necessary, can be so played back that the interrogatee is persuaded that he is the subject of the remarks. , Cooperative witnesses may be coached to exaggerate so that if a recording is played for the interrogatee or a confrontation is arranged, the source -- for example, a suspected courier -- finds the witness overstating his importance. The witness claims that the interrogatee is only incidentally a courier, that actually he is the head of an RIS kidnapping gang. The interrogator pretends amazement and says into the recorder, "I thought he was only a courier; and if he had told us the truth, I planned to let him go. But this is much more serious. On the basis of charges like these I'll have to hand him over to the local police for trial," On hearing these remarks, the interrogatee may confess the truth about the lesser guilt in order to avoid heavier punishment. If he continues to withhold, the interrogator may take his side by stating, "You know, I'm not at all convinced that so-and-so told a straight story. I feel, personally, that he was exaggerating a great deal. Wasn't he? What's the true story?"

7. Joint Suspects

If two or more interrogation sources are suspected of joint complicity in acts directed against U.S, security, they should be separated immediately. If time permits, it may be a good idea (depending upon the psychological assessment of both) to postpone interrogation for about a week. Any anxious inquiries from either can be met by a knowing grin and some such reply as, "We'll get to you in due time. There's no hurry now, " If documents, witnesses, or other sources yield information about interrogatee A, such remarks as "B says it was in Smolensk that you denounced so-and-so to the secret police. Is that right? Was it in 1937?" help to establish in A's mind the impression that B is talking. If the interrogator is quite certain of the facts in the case but cannot secure an admission from either A or B, a written confession may be prepared and A's signature may be reproduced on it. (It is helpful if B can recognize A's signature, but not essential.)

The confession contains the salient facts, but they are distorted; the confession shows that A is attempting to throw the entire responsibility upon B. Edited tape recordings which sound as though -4 had denounced B may also be used for the purpose, separately or in conjunction with the written "confession." If A is feeling a little ill or dispirited, he can also be led past a window or otherwise shown to B without creating a chance for conversation; B is likely to interpret A's hang-dog look as evidence of confession and denunciation. (It is important that in all such gambits, A be the weaker of the two, emotionally and psychologically.) B then reads (or hears) A's "confession," If B persists in withholding, the sechet interrogator should dismiss him promptly, saying that A's signed confession is sufficient for the purpose and that it does not matter whether B corroborates it or not. At the following session with B, the interrogator selects some minor matter, not substantively damaging to B but nevertheless exaggerated, and says, "I'm not sure A was really fair to you here. Would you care to tell me your side of the story?" If B rises to this bait, the interrogator moves on to areas of greater significance. The outer-and-inner office routine may also be employed. A, the weaker, is brought into the inner office, and the door is left slightly ajar or the transom open. B is later brought into the outer office by a guard and placed where he can hear, though not too clearly. The interrogator begins routine questioning of A, speaking rather softly and inducing A to follow suit. Another person in the inner office, acting by prearrangement, then quietly leads A out through another door, Any noises of departure are covered by the interrogator, who rattles the ash tray or moves a table or large chair. As soon as the second door is closed again and A is out of earshot, the interrogator resumes his questioning. His voice grows louder and angrier, He tells A to speak up, that he can hardly hear him. He grows abusive, reaches a climax, and then says, 'Well, that's better. Why didn't you say so in the first place?" The rest of the monologue is designed to give B the impression that A has now started to tell the truth. Suddenly the interrogator pops his head through the doorway and is angry on seeing B and the guard. "You jerk!" he says to the guard, 'What are you doing here?" He rides down the guard's mumbled attempt to explain the mistake, shouting, "Get him out of here! I'll take care of you later!" When, in the judgment of the interrogator, B is fairly well-convinced that A has broken down and told his story, the interrogator may elect to say to B, "Now that A has come clean with us, I'd like to let him go.

But I hate to release one of you before the other; you ought to get out at the same time. A seems to be pretty angry with you -- feels that you got him into this jam. He might even go back to your Soviet case officer and say that you haven't returned because you agreed to stay here and work for us. Wouldn't it be better for you if I set you both free together? Wouldn't it be better to tell me your side of the story?"

8. Ivan Is a Dope

It may be useful to point out to a hostile agent that the cover story was ill-contrived, that the other service botched the job, that it is typical of the other service to ignore the welfare of its agents. The interrogator may personalize this pitch by explaining that he has been impressed by the agent's courage and intelligence. He sells the agent the idea that the interrogator, not his old service, represents a true friend, who understands him and will look after his welfare.

9. Joint Interrogators

The commonest of the joint interrogator techniques is the Mutt-and-Jeff routine: the brutal, angry, domineering type contrasted with the friendly, quiet type. This routine works best with women, teenagers, and timid men. [If the interrogator who has done the bulk of the questioning up to this point has established a measure of rapport, he should play the friendly role. If rapport is absent, and especially if. antagonism has developed, the principal interrogator may take the other part. The angry interrogator speaks loudly from the beginning; and unless the interrogatee clearly indicates that he is now ready to tell his story, the angry interrogator shouts down his answers and cuts him off. He thumps the table. The quiet interrogator should not watch the show unmoved but give subtle indications that he too is somewhat afraid of his colleague. The angry interrogator accuses the subject of other offenses, any offenses, especially those that are heinous or demeaning. He makes it plain that he personally considers the interrogatee the vilest person on earth, During the harangue the friendly, quiet interrogator breaks in to say, "Wait a minute, Jim. Take it easy."' The angry interrogator shouts back, "Shut up! I'm handling this, I've broken crumb-bums before, and I'll break this one, wide open." He expresses his disgust by spitting on the floor or holding his nose or any gross gesture, Finally, red-faced and furious, he says, "I'm going to take a break, have a couple of stiff drinks. But I'll be back at two -- and you, you bum, you better be ready to talk. '' When the door slams behind him, the second interrogator tells the subject how sorry he is, how he hates to work with a man like that but has no choice, how if maybe brutes like that would keep quiet and give a man a fair chance to tell his side of the story, etc., etc. An interrogator working alone can also use the Mutt-and Jeff technique. After a number of tense and hostile sessions the interrogatee is ushered into a different or refurnished room with comfortable furniture, cigarettes, etc. The interrogator invites him to sit down and explains his regret that the source's former stubbornness forced the interrogator to use such tactics. Now everything will be different. The interrogator talks man-to-man. An American POW, debriefed on his interrogation by a hostile service that used this approach, has described the result: "Well, I went in and there was a man, an officer he was... -- he asked me to sit down and was very friendly.... It was very terrific. I, well, I almost felt like I had a friend sitting there. I had to stop every now and then and realize that this man wasn't a friend of mine....I also felt as though I couldn't be rude to him.... It was much more difficult for me to -well, I almost felt I had as much responsibility to talk to him and reason and justification as I have to talk to you right now. '(18) Another joint technique casts both interrogators in friendly roles, But whereas the interrogator in charge is sincere, the second interrogator's manner and voice convey the impression that he is merely pretending sympathy in order to trap the interrogatee. He slips in a few trick questions of the 'Whendid-you-stop-beating-your-wife?" category. The interrogator 7 in charge warns his colleague to desist. When he repeats the tactics, the interrogator in charge says, with a slight show of anger, "We're not here to trap people but to get at the truth. I suggest that you leave now. I'll handle this." It is usually unproductive to cast both interrogators in hostile roles.

Language

If the recalcitrant subject speaks more than one language, it is better to question him in the tongue with which he is least familiar as long as the purpose of interrogation is to obtain a confession. After the interrogatee admits hostile intent or activity, a switch to the better-known language will facilitate follow-up. An abrupt switch of languages may trick a resistant source, If an interrogatee has withstood a barrage of questions in German or Korean, for example, a sudden shift to "Who is your case officer?" in Russian may trigger the answer before the source can stop himself, An interrogator quite at home in the language being used may nevertheless elect to use an interpreter if the interrogatee does not know the language to be used between the interrogator and interpreter and also does not know that the interrogator knows his own tongue. The principal advantage here is that hearing everything twice helps the interrogator to note voice, expression, gestures, and other indicators more attentively. This gambit is obviously unsuitable for any form of rapid-fire questioning, and in any case it has the disadvantage of allowing the subject to pull himself together after each query. It should be used only with an interpreter who has been trained in the technique. It is of basic importance that the interrogator not using an interpreter be adept in the language selected for use. If he is not, if slips of grammar or a strong accent mar his speech, the resistant source will usually feel fortified. Almost ali people have been conditioned to relate verbal skill to intelligence, education, social status, etc. Errors or mispronunciations also permit the interrogatee to misunderstand or feign misunderstanding and thus gain time. He may also resort to polysyllabic obfuscations upon realizing the limitations of the interrogator's vocabulary.

Spinoza and Mortimer Snerd

If there is reason to suspect that a withholding source possesses useful counterintelligence information but has not had access to the upper reaches of the target organization, the policy and command level, continued questioning about lofty topics that the source knows nothing about may pave the way for the extraction of information at lower levels, The interrogatee is asked about KGB policy, for example: the relation of the service to its government, its liaison arrangements, etc., etc. His complaints that he knows nothing of such matters are met by flat insistence that he does know, he would have to know, that even the most stupid men in his position know. Communist interrogators who used this tactic against American POW's coupled it with punishment for "don't know" responses -typically by forcing the prisoner to stand at attention until he gave some positive response. After the process had been continued long enough, the source was asked a question to which he did know the answer. Numbers of Americans have mentioned ",..the tremendous feeling of relief you get when he finally asks you something you can answer," One said, ''I know it seems strange now, but I was positively grateful to them when they switched to a topic I knew something about.

The Wolf in Sheep's Clothing

It has been suggested that a successfully withholding source might be tricked into compliance if led to believe that he is dealing with the opposition. The success of the ruse depends upon a successful imitation of the opposition. A case officer previously unknown to the source and skilled in the appropriate language talks with the source under such circumstances that the latter is convinced that he is dealing with the opposition. The source is debriefed on what he has told the Americans and what he has not told them, The trick is likelier to succeed if the interrogatee has not been in confinement but a staged ' "escape, '' engineered by a stool-pigeon, might achieve the same end. Usually the trick is so complicated and risky that its employment is not recommended.

Alice in Wonderland

The aim of the Alice in Wonderland or confusion | technique is to confound the expectations and conditioned reactions of the interrogatee. He is accustomed to a world that makes some sense, at least to him: a world of continuity and logic, a predictable world. He clings to this world to reinforce his identity and powers of resistance. The confusion technique is designed not only to obliterate the familiar but to replace it with the weird. Although this method can be employed by a single interrogator, it is better adapted to use by two or three, When the subject enters the room, the first interrogator asks a double-talk question -- one which seems straightforward but is essentially nonsensical, Whether the interrogatee tries to answer or not, the second interrogator follows up (interrupting any attempted response) with a wholly unrelated and equally illogical query. Sometimes two or more questions are asked simultaneously. Pitch, tone, and volume of the interrogators' voices are unrelated to the import of the questions. No pattern of questions and answers is permitted to develop, nor do the questions themselves relate logically to each other. In this strange atmosphere the subject finds that the pattern of speech and thought which he has learned to consider normal have been replaced by an eerie meaninglessness. The interrogatee may start laughing or refuse to take the situation seriously. But as the process continues, day after day if necessary, the subject begins to try to make sense of the situation, which becomes mentally intolerable. Now he is likely to make significant admissions, or even to pour out his story, just to stop the flow of babble which assails him. This technique may be especially effective with the orderly, obstinate type.

Regression

There are a number of non-coercive techniques for inducing regression. All depend upon the interrogator's control of the environment and, as always, a proper matching of method to source, Some interrogatees can be repressed by persistent manipulation of time, by retarding and advancing clocks and serving meals at odd times -- ten minutes or ten hours after the last food was given. Day and night are jumbled. Interrogation sessions are similarly unpatterned the subject may be brought back for more questioning just a few minutes after being dismissed for the night. Half-hearted efforts to cooperate can be ignored, and conversely he can be rewarded for non-cooperation. (For example, a successfully resisting source. may become distraught if given some reward for the "valuable contribution" that he has made.)

The Alice in Wonderland technique can reinforce the effect. Two or more interrogators, questioning as a team and in relays (and thoroughly jumbling the timing of both methods) can ask questions which make it impossible for the interrogatee to give sensible, significant answers. A subject who is cut off from the world he knows seeks to recreate it, in some measure, in the new and strange environment, He may try to keep track of time, to live in the familiar past, to cling to old concepts of loyalty, to establish -- with one or more interrogators -- interpersonal relations resembling those that he has had earlier with other people, and to build other bridges back to the known. Thwarting his attempts to do so is likely to drive him deeper and deeper into himself, until he is no longer able to control his responses in adult fashion. The placebo technique is also used to induce regression.

The interrogatee is given a placebo (a harmless sugar pill). Later he is told that he has imbibed a drug, a truth serum, which will make him want to talk and which will also prevent his lying. The subject's desire to find an excuse for the compliance that represents his sole avenue of escape from his distressing predicament may make him want to believe that he has been drugged and that no one could blame him for telling his story now. Gottschelk observes, "Individuals under increased stress are more likely to respond to placebos. '(7) Orne has discussed an extension of the placebo concept in explaining what he terms the "magic room" technique. 'An example. . . would be. . . the prisoner who is given a hypnotic suggestion that his hand is growing warm. However in this instance, the prisoner's hand actually does become warm, a problem easily resolved by the use of a concealed diathermy machine, Or it might be suggested...that...a cigarette will taste bitter, Here again, he could be given a cigarette prepared to have a slight but noticeably bitter taste, "' In discussing states of heightened suggestibility (which are not, however, states of trance) Orne says, "Both hypnosis and some of the drugs inducing hypnoidal states are Popularly viewed as situations where the individual is no longer master of his own fate and therefore not responsible for his actions, It seems possible then that the hypnotic situation, as distinguished from hypnosis itself, might be used to relieve the individual of a feeling of responsibility for his own actions and thus lead him to reveal information. '"{7) In other words, a psychologically immature source, or one who has been regressed, could adopt an implication or suggestion that he has been drugged, hypnotized, or otherwise rendered incapable of resistance, even if he recognizes at some level that the suggestion is untrue, because of his strong desire to escape the stress of the situation by capitulating. These techniques provide the source with the rationalization that he needs. Whether regression occurs spontaneously under detention or interrogation, and whether it is induced by a coercive or non-coercive technique, it should not be allowed to continue past the point necessary to obtain compliance. Severe techniques of regression are best employed in the presence of a psychiatrist, to insure full reversal later. As soon as he can, the interrogator presents the subject with the way out, the face-saving reason for escaping from his painful dilemma by yielding. Now the interrogator becomes fatherly. Whether the excuse is that others have already confessed ("all the other boys are doing it"), that the interrogatee has a chance to redeem himself ('you're really a good boy at heart"), or that he can't help himself ("they made you do it"), the effective rationalization, the one the source will jump at, is likely to be elementary. It is an adult's version of the excuses of childhood.

The Polygraph

The polygraph can be used for purposes other than the evaluation of veracity. For example, it may be used as an adjunct in testing the range of languages spoken by an interrogatee or his sophistication in intelligence matters, for rapid screening to determine broad areas of knowledgeability, and as an aid in the psychological assessment of sources. Its primary function in a counterintelligence interrogation, however, is to provide a further means of testing for deception or withholding. A resistant source suspected of association with a hostile clandestine organization should be tested polygraphically at least once. Several examinations may be needed. As a general rule, the polygraph should not be employed as a measure of last resort. More reliable readings will be obtained if the instrument is used before the subject has been placed under intense pressure, whether such pressure is coercive or not. Sufficient information for the purpose is normally available after screening and one or two interrogation sessions. Although the polygraph has been a valuable aid, no interrogator should feel that it can carry his responsibility for him, "The polygraph lays no claim to one-hundred-percent reliability. Test results can be as varied as the individuals tested, and the interpretation of the charts is not a simple matter of deciding whether the subject reacted or did not react. Many charts are quite definitive; but some indicate only a probability and from two to five percent of the cases tested end up being classified as inconclusive, with crucial areas left unresolved, '(9) The best results are obtained when the CI interrogator ' and the polygraph operator work closely together in laying the groundwork for technical examination. The operator needs all available information about the personality of the source, as well as the operational background and reasons for suspicion. The Cl interrogator in turn can cooperate more effectively and can fit the results of technical examination more accurately into the totality of his findings, The following discussion is based upon R,C, Davis' "Physiological Responses as a Means of Evaluating Information. " (7) Although improvements appear to be in the offing, the instrument in widespread use today measures breathing, systolic blood pressure, and galvanic skin response (GSR). "One drawback in the use of respiration as an indicator, " according to Davis, "is its susceptibility to voluntary control." Moreover, if the source "knows that changes in breathing will disturb all physiologic variables under control of the autonomic division of the nervous system, and possibly even some others, a certain amount of cooperation or a certain degree of ignorance is required for lie detection by physiologic methods to work." In general, ". . . breathing during deception is shallower and slower than in truth telling. . . the inhibition of breathing seems rather characteristic of anticipation of a stimulus." The measurement of systolic blood pressure provides a reading on a phenomenon not usually subject to voluntary control. The pressure ". . . will typically rise by a few millimeters of mercury in response to a question, whether it is answered truthfully or not,

The evidence is that the rise will generally be greater when (the subject) is lying. " However, discrimination between truth-telling and lying on the basis of both breathing and blood pressure ", , . is poor (almost nil) in the early part of the sitting and improves to a high point later. " The galvanic skin response is one of the most easily triggered reactions, but recovery after the reaction is slow, and", , , in a routine examination the next question is likely to be introduced before recovery is complete. Partly because of this fact there is an adapting trend in the GSR; with stimuli repeated every few minutes the response gets smaller, other things being equal." Davis examines three theories regarding the polygraph. The conditional response theory holds that the subject reacts to questions that strike sensitive areas, regardless of whether he is telling the truth or not. Experimentation has not substantiated this theory. The theory of conflict presumes that a large physiologic disturbance occurs when the subject is caught between his habitual inclination to tell the truth and his strong desire not to divulge a certain set of facts. Davis suggests that if this concept is valid, it holds only if the conflict is intense. The threat-of-punishment theory maintains that a large physiologic response accompanies lying because the subject fears the consequence of failing to deceive, 'In common language it might be said that he fails to deceive the machine operator for the very reason that he fears he will fail, The 'fear' would be the very reaction detected," This third theory is more widely held than the other two, Interrogators should note the inference that a resistant source who does not fear that detection of lying will result in a punishment of which he is afraid would not, according to this theory, produce significant responses.

Graphology

The validity of graphological techniques for the analysis of the personalities of resistant interrogatees has not been established. There is some evidence that graphology is a useful aid in the early detection of cancer and of certain mental illnesses. If the interrogator or his unit decides to have a source's handwriting analyzed, the samples should be submitted to Headquarters as soon as possible, because the analysis is more useful in the preliminary assessment of the source than in the later interrogation, Graphology does have the advantage of being one of the very few techniques not requiring the assistance or even the awareness of the interrogatee. As with any other aid. the interrogator is free to determine for himself whether the analysis provides him with new and valid insights, confirms other observations, is not helpful, or is misleading.

IX. THE COERCIVE COUNTERINTELLIGENCE INTER-ROGATION OF RESISTANT SOURCES

A. Restrictions

The purpose of this part of the handbook is to present basic information about coercive techniques available for use in the interrogation situation. It is vital that this discussion not be misconstrued as constituting authorization for the use of coercion at field discretion. As was noted earlier, there is no such blanket authorization. Prior Headquarters approval t the KUDOVE level must be obtained for the interrogation of from national against his will under any of the following circumstances: (1) if bodily harm is to be inflicted; (2) if medical, chemical, or electrical methods or materials are to be used to induce an acquiescence the detention is locally illegal and traceable: except that in cases of extreme operational urgency requiring immediate detention, retroactive Headquarters approval may be promptly requested by priority cable, For both ethical and pragmatic reasons no interrogator may take upon himself the unilateral responsibility for using coercive methods. Concealing from the interrogator's superiors an intent to resort to coercion, or its unapproved employment, does not protect them. It places them, and KUBARK, in unconsidered jeopardy.

B. The Theory of Coercion

Coercive procedures are designed not only to exploit the resistant source's internal conflicts and induce him to wrestle with himself but also to bring a superior outside force to bear upon the subject's resistance. Non-coercive methods are not likely to succeed if their selection and use is not predicated upon an accurate psychological assessment of the source. In contrast, the same coercive method may succeed against persons who are very unlike each other. The changes of success rise steeply, nevertheless, if the coercive technique is matched to the source's personality. Individuals react differently even to such seemingly non-discriminatory stimuli as drugs. Moreover, it is a waste of time and energy to apply strong pressures on a hit-or-miss basis if a tap on the psychological jugular will produce compliance. All coercive techniques are designed to induce regression. As Hinkle notes in "The Physiological State of the Interrogation Subject as it Affects Brain Function"(7), the result of external pressures of sufficient intensity is the loss of those defenses most recently acquired by civilized man: ". . . the capacity to carry out the highest creative activities, to meet new, challenging, and complex situations, to deal with trying interpersonal relations, and to cope with repeated frustrations. Relatively small degrees of homeostatic derangement, fatigue, pain, sleep loss, or anxiety may impair these functions." As a result, "most people who are exposed to coercive procedures will talk and usually reveal some information that they might not have revealed otherwise. '' One subjective reaction often evoked by coercion is a feeling of guilt. Meltzer observes, "In some lengthy interrogations, the interrogator may, by virtue of his role as the sole supplier of satisfaction and punishment, assume the stature and importance of a parental figure in the prisoner's feeling and thinking. Although there may be intense hatred for the interrogator, it is not unusual for warm feelings also to develop.

This ambivalence is the basis for guilt reactions, and if the interrogator nourishes these feelings, the guilt may be strong enough to influence the prisoner's behavior... . Guilt makes compliance more likely. . . ." (7). Farber says that the response to coercion typically contains ". . . at least three important elements: debility, dependency, and dread." Prisoners". . . have reduced viability, are helplessly dependent their captors for the satisfaction of their many basic needs, and experience the emotional and motivational reactions of intense fear and anxiety. . . . Among the /American/ POW's pressured by the Chinese Communists, the DDD syndrome in its full-blown form constituted a state of discomfort that was well-nigh intolerable." (ll). If the debility-dependency~dread state is unduly prolonged, however, the arrestee may sink into a defensive apathy from which it is hard to arouse him. Psychologists and others who write about physical or psychological duress frequently object that under sufficient pressure subjects usually yield but that their ability to recall and communicate information accurately is as impaired as the will to resist. This pragmatic objection has somewhat the same validity for a counterintelligence interrogation as for any other. But there is one significant difference. Confession is a necessary prelude to the CI interrogation of a hitherto unresponsive or concealing source. And the use of coercive techniques will rarely or never confuse an interrogatee so completely that he does not know whether his own confession is true or false. He does not need full mastery of all his powers of resistance and discrimination to know whether he is a spy or not. Only subjects who have reached a point where they are under delusions are likely to make false confessions that they believe. Once a true confession is obtained, the classic cautions apply. The pressures are lifted, at least enough so that the subject can provide counterintelligence information as accurately as possible. In fact, the relief granted the subject at this time fits neatly into the interrogation plan. He is told that the changed treatment is a reward for truthfulness and an evidence that friendly handling will continue as long as he cooperates. The profound moral objection to applying duress past the point of irreversible psychological damage has been stated. Judging the validity of other ethical arguments about coercion exceeds the scope of this paper. What is fully clear, however, is that controlled coercive manipulation of an interrogatee may impair his ability to make fine distinctions but will not alter his ability to answer correctly such gross questions as "Are you a Soviet agent? What is your assignment now? Who is your present case officer?"

When an interrogator senses that the subject's resistance is wavering, that his desire to yield is growing stronger than his wish to continue his resistance, the time has come to provide him with the acceptable rationalization: a face-saving reason or excuse for compliance, Novice interrogators may be tempted to seize upon the initial yielding triumphantly and to personalize the victory. Such a temptation must be rejected immediately. An interrogation is not a game played by two people, one to become the winner and the other the loser. It is simply a method of obtaining correct and useful information. Therefore the interrogator should intensify the subject's desire to cease struggling by showing him how he can do so without seeming to abandon principle, self-protection, or other initial causes of resistance.

If, instead of providing the right rationalization at the right time, the interrogator seizes gloatingly upon the subject's wavering, opposition will stiffen again, The following are the principal coercive techniques of interrogation: arrest, detention, deprivation of sensory stimuli through solitary confinement or similar methods, threats and fear, debility, pain, heightened suggestibility and hypnosis, narcosis, and induced regression, This section also discusses the detection of malingering by interrogatees and the provision of appropriate rationalizations for capitulating and cooperating.

C. Arrest

The manner and timing of arrest can contribute substantially to the interrogator's purposes. 'What we aim to do is to ensure that the manner of arrest achieves, if possible, surprise, and the maximum amount of mental discomfort in order to catch the suspect off balance and to deprive him of the initiative. One should therefore arrest him at a moment when he least expects it and when his mental and physical resistance is at its lowest, The ideal time at which to arrest a person is in the early hours of the morning because surprise is achieved then, and because : a person's resistance physiologically as well as psychologically ; is at its lowest.... If a person cannot be arrested in the early hours..., then the next best time is in the evening.... "Then, .as to the nature of arrest, it is of great importance that the arresting parties . . . behave in such a manner as to impress the suspect with their efficiency. ... If the suspect... sees three or four ill-dressed, ill-equipped, slovenly policemen, he is more likely to recover from the initial shock, and to think that he has fallen into the hands of persons whom he might easily be able to outwit. If, however, he is rudely awakened by an arresting party of particularly large, particularly smart, particularly well-equipped, particularly efficient policemen, he will probably become exceedingly depressed and anxious about his future." (1)

D. Detention

If, through the cooperation of a liaison service Gr by unilateral means,/arrangements have been made for the confinement of a resistant source, the circumstances of detention are arranged to enhance within the subject his feelings of being cut off from the known and the reassuring, and of being plunged into the strange. Usually his own clothes are immediately taken away, because familiar clothing reinforces identity and thus the capacity for resistance. (Prisons give close hair cuts and issue prison garb for the same reason.) If the interrogatee is especially proud or neat, it may be useful to give him an outfit that is one or two sizes too large and to fail to provide a belt, so that he must hold his pants up. The point is that man's sense of identity depends upon a continuity in his surroundings, habits, appearance, actions, relations with others, etc. Detention permits the interrogator to cut through these links and throw the interrogatee back upon his own unaided internal resources. Little is gained if confinement merely replaces one routine with another. Prisoners who lead monotonously unvaried lives "... cease to care about their utterances, dress, and cleanliness. They become dulled, apathetic, and depressed."" (7) And apathy can be a very effective defense against interrogation. Control of the source's environment permits the interrogator to determine his diet, sleep pattern, and other fundamentals. Manipulating these into irregularities, so that the subject becomes disorientated, is very likely to create feelings of fear and helplessness. Hinkle points out, "People who enter prison with attitudes of foreboding, apprehension, and helplessness generally do less well than those who enter with assurance and a conviction that they can deal with anything that they may encounter.... . Some people who are afraid of losing sleep, or who do not wish to lose sleep, soon succumb to sleep loss ... .' (7) In short, the prisoner should not be provided a routine to which he can adapt and from which he can draw some comfort-or at least a sense of his own identity. Everyone has read of prisoners who were reluctant to leave their cells after prolonged incarceration. Little is known about the duration of confinement calculated to make a subject shift from anxiety, coupled with a desire for sensory stimuli and human companionship, to a passive, apathetic acceptance of isolation and an ultimate pleasure in this negative state. Undoubtedly the rate of change is determined almost entirely by the psychological characteristics of the individual. In any event, it is advisable to keep the subject upset by constant disruptions of patterns.

For this reason, it is useful to determine whether the interrogattee has been jailed before, how often, under what circumstances, for how long, and whether he was subjected to earlier interrogation. Familiarity with confinement and even with isolation reduces the effect.

E. Deprivation of Sensory Stimuli

The chief effect of arrest and detention, and particularly of 5 solitary confinement, is to deprive the subject of many or most of the sights, sounds, tastes, smells, and tactile sensations to which he has grown accustomed.

John C. Lilly examined eighteen autobiographical accounts written by polar explorers and solitary seafarers. He found". .. that isolation per se acts on most persons as a powerful stress.... . In all cases of survivors of isolation at sea or in the polar night, it was the first exposure which caused the greatest fears and hence the greatest danger of giving way to symptoms; previous experience is a powerful aid in going © ahead, despite the symptoms. "The symptoms most commonly produced by isolation are superstition, intense love of any other living thing, perceiving inanimate objects as alive, hallucinations, and delusions." (26) The apparent reason for these effects is that a person cut off from external stimuli turns his awareness inward, upon himself, and then projects the contents of his own unconscious outwards, so that he endows his faceless environment with his own attributes, fears, and forgotten memories, Lilly notes, "It is obvious that inner factors in the mind tend to be projected outward, that some of the mind's activity which is usually reality bound now becomes free to turn to phantasy and ultimately to hallucination and delusion, " A number of experiments conducted at McGill University, the National Institute of Mental Health, and other sites have attempted to come as close as possible to the elimination of sensory stimuli, or to masking remaining stimuli, chiefly sounds, by a stronger but wholly monotonous overlay. The results of these experiments have little applicability to interrogation because the circumstances are dissimilar. Some of the findings point toward hypotheses that seem relevant to interrogation, but conditions like those of detention for purposes of counterintelligence interrogation have not been duplicated for experimentation.

At the National Institute of Mental Health two subjects were ", , . suspended with the body and all but the top of the head immersed in a tank containing slowly flowing water at 34.5°C (94.5° F). . . ." Both subjects wore black-out masks, which enclosed the whole head but allowed breathing and nothing else. The sound level was extremely low; the subject heard only his own breathing and some faint sounds of water from the piping. Neither subject stayed in the tank longer than three hours, Both passed quickly from normally directed thinking through a tension resulting from unsatisfied hunger for sensory stimuli and concentration upon the few available sensations to private reveries and fantasies and eventually to visual imagery somewhat resembling hallucinations. "In our experiments, we notice that after immersion the day apparently is started over, i.e., the subject feels as if he has risen from bed afresh; this effect persists, and the subject finds he is out of step with the clock for the rest of the day."

Drs. Wexler, Mendelson, Leiderman, and Solomon conducted a somewhat similar experiment on seventeen paid volunteers. These subjects were '""... placed in a tank-type respirator with a specially built mattress.... The vents of the respirator were left open, so that the subject breathed for himself. His arms and legs were enclosed in comfortable but rigid cylinders to inhibit movement and tactile contact. The subject lay on his back and was unable to see any part of his body. The motor of the respirator was run constantly, producing a dull, repetitive auditory stimulus. The room admitted no natural light, and artificial light was minimal and constant." (42) Although the established time limit was 36 hours and though all physical needs were taken care of, only 6 of the 17 completed the stint. The other eleven soon asked for release. Four of these terminated the experiment because of anxiety and panic; seven did so because of physical discomfort. The-results confirmed earlier findings that (1) the deprivation of sensory stimuli induces stress; (2) the stress becomes unbearable for most subjects; (3) the subject has a growing need for physical and social stimuli; and (4) some subjects progressively lose touch with reality, focus inwardly, and produce delusions, hallucinations, and other pathological effects. In summarizing some scientific reporting on sensory and perceptual deprivation, Kubzansky offers the following observations: "Three studies suggest that the more well-adjusted or 'normal' the subject is, the more he is affected by deprivation of sensory stimuli. Neurotic and psychotic subjects are either comparatively unaffected or show decreases in anxiety, hallucinations, etc."

These findings suggest - but by no means prove - the following theories about solitary confinement and isolation:

1. The more completely the place of confinement eliminates sensory stimuli, the more rapidly and deeply will the interrogatee be affected. Results produced only after weeks or months of imprisonment in an ordinary cell can be duplicated in hours or days in a cell which has no light (or weak artificial light which never varies), which is sound-proceed, in which odors are eliminated, etc. An environment still more subject to control, such as water-tank or iron lung, is even more effective.

2. An early effect of such an environment is anxiety. How soon it appears and how strong it is depends upon the psychological characteristics of the individual.

3. The interrogator can benefit from the subject's anxiety. As the interrogator becomes linked in the subject's mind with the reward of lessened anxiety, human contact, and meaningful activity, and thus with providing relief for growing discomfort, the questioner assumes a benevolent role. (7)

4. The deprivation of stimuli induces regression by depriving the subject's mind of contact with an outer world and thus forcing it in upon itself. At the same time, the calculated provision of stimuli during interrogation tends to make the regressed subject view the interrogator as a father figure. The result, normally, is a strengthening of the subject's tendencies toward compliance.

F. Threats and Fear

The threat of coercion usually weakens or destroys resistance more effectively than coercion itself. The threat to inflict pain, for example, can trigger fears more damaging than the immediate sensation of pain. In fact, most people underestimate their capacity to withstand pain. The same principle holds for other fears: sustained long enough, a strong fear of anything vague or unknown induces regression, whereas the materialization of the fear, the infliction of some form of punishment, is likely to come as a relief. The subject finds that he can hold out, and his resistances are strengthened. © "In general, direct physical brutality creates only resentment, hostility, and further defiance." (18) The effectiveness of a threat depends not only on what sort of person the interrogatee is and whether he believes that his questioner can and will carry the threat out but also on the interrogator's reasons for threatening. If the interrogator threatens because he is angry, the subject frequently senses the fear of failure underlying the anger and is strengthened in his own resolve to resist. Threats delivered coldly are more effective than those shouted in rage. It is especially important that a threat not be uttered in response to the interrogatee's own expressions of hostility. These, if ignored, can induce feelings of guilt, whereas retorts in kind relieve the subject's feelings. Another reason why threats induce compliance not evoked by the inflection of duress is that the threat grants the interrogatee time for compliance. It is not enough that a resistant source should be placed under the tension of fear; he must also discern an acceptable escape route. Biderman observes, "Not only can the shame or guilt of defeat in the encounter with the interrogator be involved, but also the more fundamental injunction to protect one's self-autonomy or 'will'.... A simple defense against threats to the self from the anticipation of being forced to comply is, of course, to comply 'deliberately' or' voluntarily'.... To the extent that the foregoing interpretation holds, the more intensely motivated the interrogatee is to resist, the more intense is the , pressure toward early compliance from such anxieties, for the greater is the threat to self-esteem which is involved in contemplating the possibility of being 'forced to' comply' (6)

In brief, the threat is like all other coercive ' techniques in being most effective when so used as to foster regression and when joined with a suggested way out of the dilemma, a rationalization acceptable to the interrogatee.

The threat of death has often been found to be worse than useless. It "has the highest position in law as a defense, but in many interrogation situations it is a highly ineffective threat. Many prisoners, in fact, have refused to yield in the face of such threats who have subsequently been 'broken' by other procedures." (3) The principal reason is that the ultimate threat is likely to induce sheer hopelessness if the interrogatee does not believe that it is a trick; he feels that he is as likely to be condemned after compliance as before. The threat of death is also ineffective when used against hard-headed types who realize that silencing them forever would defeat the interrogator's purpose. If the threat is recognized as a bluff, it will not only fail but also pave the way to failure for later coercive ruses used by the interrogator.

G. Debility

No report of scientific investigation of the effect of debility upon the interrogatee's powers of resistance has been discovered. For centuries interrogators have employed various methods of inducing physical weakness: prolonged constraint; prolonged exertion; extremes of heat, cold, or moisture; and deprivation or drastic reduction of food or sleep.

Apparently the assumption is that lowering the source's physiological resistance will lower his psychological capacity for opposition. If this notion were valid, however, it might reasonably be expected that those subjects who are physically weakest at the beginning of an interrogation would be the quickest to capitulate, a concept not supported by experience. The available evidence suggests that resistance is sapped principally by psychological rather than physical pressures. The threat of debility - for example, a brief deprivation of food - may induce much more anxiety than prolonged hunger, which will result after a while in apathy and, perhaps, eventual delusions or hallucinations. In brief, it appears probable that the techniques of inducing debility become counterproductive at an early stage.

The discomfort, tension, and restless search for an avenue of escape are followed by withdrawal symptoms, a turning away from external stimuli, and a sluggish unresponsiveness. Another objection to the deliberate inducing of debility is that prolonged exertion, loss of sleep, etc., themselves become patterns to which the subject adjusts through apathy. The interrogator should use his power over the resistant subject's physical environment to disrupt patterns of response, not to create them. Meals and sleep granted irregularly, in more than abundance or less than adequacy, the shifts occurring on no discernible time pattern, will normally disorient an interrogatee and sap his will to resist more effectively than a sustained deprivation leading to debility.

H. Pain

Everyone is aware that people react very differently to pain. The reason, apparently, is not a physical difference in the intensity of the sensation itself. Lawrence E. Hinkle observes, "The sensation of pain seems to be roughly equal in all men, that is to say, all people have approximately the same threshold at which they begin to feel pain, and when carefully graded stimuli are applied to them, their estimates of severity are approximately the same.... Yet... when men are very highly motivated...they have been known to carry out rather complex tasks while enduring the most intense pain." He also states, "In general, it appears that whatever may be the role of the constitutional endowment in determining the reaction to pain, it is a much less j important determinant than is the attitude of the man who experiences the pain." (7) The wide range of individual reactions to pain may be partially explicable in terms of early conditioning.

The person whose first encounters with pain were frightening and intense may be more violently affected by its later infliction than one whose original experiences were mild. Or the reverse may be true, and the man whose childhood familiarized him with pain may dread it less, and react less, than one whose distress is heightened by fear of the unknown. The individual remains the determinant. It has been plausibly suggested that, whereas pain inflicted on a person from outside himself may actually focus or intensify his will to resist, his resistance is likelier to be sapped by pain which he seems to inflict upon himself. "In the simple torture situation the contest is one between the individual and his tormentor (.... and he can frequently endure). When the individual is told to stand at attention for long periods, an intervening factor is introduced. The immediate source of pain is not the interrogator but the victim himself.

The motivational strength of the individual is likely to exhaust itself in this internal encounter.... As long as the subject remains standing, he is attributing to his captor the power to do something worse to him, but there is actually no showdown of the ability of the interrogator to do so." (4) Interrogatees who are withholding but who feel qualms of guilt and a secret desire to yield are likely to become intractable if made to endure pain. The reason is that they can then interpret the pain as punishment and hence as expiation. There are also persons who enjoy pain and its anticipation and who will keep back information that they might otherwise divulge if they are given reason to expect that withholding will result in the punishment that they want. Persons of considerable moral or intellectual stature often find in pain inflicted by others a confirmation of the belief that they are in the hands of inferiors, and their resolve not to submit is strengthened. Intense pain is quite likely to produce false confessions, concocted as a means of escaping from distress. A time consuming delay results, while investigation is conducted and the admissions are proven untrue. During this respite the interrogatée can pull himself together. He may even use the time to think up new, more complex "admissions" that take still longer to disprove. KUBARK is especially vulnerable to such tactics because the interrogation is conducted for the sake of information and not for police purposes.

If an interrogatee is caused to suffer pain rather late in the interrogation process and after other tactics have failed, he is almost certain to conclude that the interrogator is becoming desperate. He may then decide that if he can just hold out against this final assault, he will win the struggle and his freedom. And he is likely to be right. Interrogatees who have withstood pain are more difficult to handle by other methods. The effect has been not to repress the subject but to restore his confidence and maturity.

I. Heightened Suggestibility and Hypnosis

In recent years a number of hypotheses about hypnosis have been advanced by psychologists and others in the guise of proven principles. Among these are the flat assertions that a person connot be hypnotized against his will; that while hypnotized he cannot be induced to divulge information that he wants urgently to conceal; and that he will not undertake, in trance or through post-hypnotic suggestion, actions to which he would normally have serious moral or ethical objections. If these and related contentions were proven valid, hypnosis would have scant value for the interrogator. But despite the fact that hypnosis has been an object of scientific inquiry for a very long time, none of these theories has yet been tested adequately. Each of them is in conflict with some observations of fact. In any event, an interrogation handbook cannot and need not include a lengthy discussion of hypnosis. The case officer or interrogator needs to know enough about the subject to understand the circumstances under which hypnosis can be a useful tool, so that he can request expert assistance appropriately. Operational personnel, including interrogators, who chance to have some lay experience or skill in hypnotism should not themselves use hypnotic techniques for interrogation or other operational purposes. There are two reasons for this position. The first is that hypnotism used as an operational tool by a practitioner who is not a psychologist, psychiatrist, or M.D. can produce irreversible psychological damage.

The lay practitioner does not know enough to use the technique safely. The second reason is that an unsuccessful attempt to hypnotize a subject for purposes of interrogation, or a successful attempt not adequately covered by post-hypnotic amnesia or other protection, can easily lead to lurid and embarrassing publicity or legal charges. Hypnosis is frequently called a state of heightened suggestibility, but the phrase is a description rather than a definition. Merton M. Gill and Margaret Brenman state, "The psychoanalytic theory of hypnosis clearly implies, where it does not explicitly state, that hypnosis is a form of regression." And they add, "...induction/f hypnosis/ is the process of bringing about a regression, while the hypnotic state is the established regression." (13) It is suggested that the interrogator will find this definition the most useful. The problem of overcoming the resistance of an uncooperative interrogatee is essentially a problem of inducing regression to a level at which the resistance can no longer be sustained. Hypnosis is one way of regressing people. Martin T. Orne has written at some length about hypnosis and interrogation. Almost all of his conclusions are tentatively negative. Concerning the role played by the will or attitude of the interrogatee, Orne says, "Although the crucial experiment has not yet been done, there is little or no evidence to indicate that trance can be induced against a person's wishes." He adds, "...the actual occurrence of the trance state is related to the wish of the subject to enter hypnosis."

And he also observes, "...whether a subject will or will not enter trance depends upon his relationship with the hypnotist rather than upon the technical procedure of trance induction."" These views are probably representative of those of many psychologists, but they are not definitive. As Orne himself later points out, the interrogatee "...could be given a hypnotic drug with appropriate verbal suggestions to talk about a given topic. Eventually enough of the drug 96 sachet would be given to cause a short period of unconsciousness, When the subject wakesn, the interrogator could then read from his 'notes' of the hypnotic interview the information presumably told him." (Orne had previously pointed out that this technique requires that the interrogator possess significant information about the subject without the subject's knowledge.) "It can readily be seen how this... maneuver... would facilitate the elicitation of information in subsequent interviews." (7) Techniques of inducing trance in resistant subjects through preliminary administration of so-called silent drugs (drugs which the subject does not know he has 'taken) or through other non-routine methods of induction are still under investigation. Until more facts are known, the question of whether a resister can be hypnotized involuntarily must go unanswered, Orne also holds that even if a resister can be hypnotized, his resistance does not cease. He postulates "... that only in rare interrogation subjects would a sufficiently deep trance be obtainable to even attempt to induce the subject to discuss material which he is unwilling to discuss in the waking state. The kind of information which can be obtained in these rare instances is still an unanswered question. '"" He adds that it is doubtful that a subject in trance could be made to reveal information which he wished to safeguard. But here too Orne seems somewhat too cautious or pessimistic. Once an interrogatee is in a hypnotic trance, his understanding of reality becomes subject to manipulation. For example, a KUBARK interrogator could tell a suspect double agent in trance that the KGB is conducting the questioning, and thus invert the whole frame of reference. In other words, Orne is probably right in holding that most recalcitrant subjects will continue effective resistance as long as the frame of reference is undisturbed. But once the subject is tricked into believing that he is talking to friend rather than foe, or that divulging the truth is the best way to serve his own purposes, his resistance will be replaced by cooperation. The value of hypnotic trance is not that it permits the interrogator to impose his will but rather that it can be used to convince the interrogatee that there is no valid reason not to be forthcoming.

A third objection raised by Orne and others is that material elicited during trance is not reliable. Orne says, "...it has been shown that the accuracy of such information... would not be guaranteed since subjects in hypnosis are fully capable of lying.'"" Again, the observation is correct; no known manipulative method guarantees veracity. But if hypnosis is employed not as an immediate instrument for digging out the truth but rather as a way of making the subject want to align himself with his interrogators, the objection evaporates. Hypnosis offers one advantage not inherent in other interrogation techniques or aids: the post-hypnotic suggestion. Under favorable circumstances it should be possible to administer a silent drug to a resistant source, persuade him as the drug takes effect that he is slipping into a hypnotic trance, place him under actual hypnosis as consciousness is returning, shift his frame of reference so that his reasons for resistance become reasons for cooperating, interrogate him, and conclude the session by implanting the suggestion that when he emerges from trance he will not remember anything about what has happened. This sketchy outline of possible uses of hypnosis in the interrogation of resistant sources has no higher goal than to remind operational personnel that the technique may provide the answer to a problem not otherwise soluble. To repeat: hypnosis is distinctly not a do-it-yourself project. Therefore the interrogator, base, or center that is considering its use must anticipate the timing sufficiently not only to secure the obligatory headquarters permission but also to allow for an expert's travel time and briefing.

J. Narcosis

Just as the threat of pain may more effectively induce compliance than its infliction, so an interrogatee's mistaken belief that he has been drugged may make him a more useful interrogation subject than he would be under narcosis. Louis A. Gottschalk cites a group of studies as indicating "that 30 to 50 per cent of individuals are placebo reactors, that is, respond with symptomatic relief to taking an inert substance." (7) In the interrogation situation, moreover, the effectiveness of a placebo may be enhanced because of its ability to placate the conscience. The subject's primary source of resistance to confession or divulgence may be pride, patriotism, personal loyalty to superiors, or fear of retribution if he is returned to their hands. Under such circumstances his natural desire to escape from stress by complying with the interrogator's wishes may become decisive if he is provided _an acceptable rationalization for compliance.

"I was drugged" is one of the best excuses. Drugs are no more the answer to the interrogator's prayer than the polygraph, 'hypnosis, or other aids. Studies and reports "dealing with the validity of material extracted from reluctant informants. ..indicate that there is no drug which can force every informant to report all the information he has. Not only may the inveterate criminal psychopath lie under the influence of drugs which have been tested, but the relatively normal and well-adjusted individual may also successfully disguise factual data.'" (3) Gottschalk reinforces the latter observation in mentioning an experiment involving drugs which indicated that "the more normal, well-integrated individuals could lie better than the guilt-ridden, neurotic subjects." (7) Nevertheless, drugs can be effective in overcoming resistance not dissolved by other techniques. As has already been noted, the so-called silent drug (a pharmacologically potent substance given to a person unaware of its administration) can make possible the induction of hypnotic trance ina previously unwilling subject. Gottschalk says, "The judicious choice of a drug with minimal side effects, its matching to the subject's personality, careful gauging of dosage, and a sense of timing.../ make/ silent administration a hard-to-equal ' ally for the hypnotist intent on producing self-fulfilling and inescapable suggestions... the drug effects should prove... compelling to the subject since the perceived sensations originate entirely within himself." (Particularly important is the reference to matching the drug to the personality of the interrogatee. The effect of most drugs depends more upon the personality of the subject than upon the physical characteristics of the drugs themselves. If the approval of Headquarters has been obtained and if a doctor is at hand for administration, one of the most important of the interrogator's functions is providing the doctor with a full and accurate description of the psychological make-up of the interrogatee, to facilitate the best possible choice of a drug. Persons burdened with feelings of shame or guilt are likely to unburden themselves when drugged, especially if these feelings have been reinforced by the interrogator. And like the placebo, the drug provides an excellent rationalization of helplessness for the interrogatee who wants to yield but has hitherto been unable to violate his own values or loyalties. Like other coercive media, drugs may affect the content of what an interrogatee divulges. Gottschalk notes that certain drugs "may give rise to psychotic manifestations such as hallucinations, illusions, delusions, or disorientation", so that "the verbal material obtained cannot always be considered valid." (7) For this reason drugs (and the other aids discussed in this section) should not be used persistently to facilitate the inter-rogative debriefing that follows capitulation. Their function is to cause capitulation, to aid in the shift from resistance to coopera-tion. Once this shift has been accomplished, coercive techniques should be abandoned both for moral reasons and because they are unnecessary and even counter-productive. This discussion does not include a list of drugs that have been employed for interroga-tion purposes or a discussion of their properties because these are medical considerations within the province of a doctor rather than an interogator.

K. The Detection of Malingering

The detection of malingering is obviously not an interrogation technique, coercive or otherwise. But the history of interrogation is studded with the stories of persons who have attempted, often successfully, to evade the mounting pressures of interrogation by feigning physical or mental illness. KUBARK interrogators may encounter seemingly sick or irrational interrogatees at times and places which make it difficult or next-to-impossible to summon medical or other professional assistance. Because a few tips may make it possible for the interrogator to distinguish between the malingerer and the person who is genuinely ill, and because both illness and malingering are sometimes produced by coercive inter-rogation, a brief discussion of the topic has been included here. Most persons who feign a mental or physical illness itn do not know enough about it to deceive the well-informed. Malcolm L. Meltzer says, "The detection of malingering depends to a great extent on the simulator's failure to understand adequately the characteristics of the role he is feigning.... Often he presents symptoms which are exceedingly rare, existing mainly in the fancy of the layman. One such symptom is the delusion of misidentification, characterized by the...belief that he is some powerful or historic personage. This symptom is very unusual in true psychosis, but is used by a number of simulators. In schizophrenia, the onset tends to be gradual, delusions do not spring up full-blown over night; in simulated disorders, the onset is usually fast and delusions may be readily available. The feigned psychosis often contains many contradictory and inconsistent symptoms, rarely existing together. The malingerer tends to go to extremes in his portrayal of his symptoms; he exaggerates, overdramatizes, grimaces, shouts, is overly bizarre, and calls attention to himself in other ways.... "Another characteristic of the malingerer is that he will usually seek to evade or postpone examination. A study of the behavior of lie-detector subjects, for example, showed that persons later 'proven guilty' showed certain similarities of behavior. The guilty persons were reluctant to take the test, and they tried in various ways to postpone or delay it.

They often appeared highly anxious and sometimes took a hostile attitude toward the test and the examiner. Evasive tactics sometimes appeared, such as sighing, yawning, moving about, all of which foil the examiner by obscuring the recording. Before the examination, they felt it necessary to explain why their responses might mislead the examiner into thinking they were lying. Thus the procedure of subjecting a suspected -malingerer to a lie-detector test might evoke behavior which would reinforce the suspicion of fraud." (7) Meltzer also notes that malingerers who are not professional psychologists can usually be exposed through Rorschach tests. An important element in malingering is the frame of mind of the examiner. A person pretending madness awakens in a professional examiner not only suspicion but also a desire to expose the fraud, whereas a well person who pretends to be concealing mental illness and who permits only a minor symptom or two to peep through is much likelier to create in the expert a desire to expose the hidden sickness, Meltzer observes that simulated mutism and amnesia can usually be distinguished from the true states by narcoanalysis. The reason, however, is the reverse of the popular misconception. Under the influence of appropriate drugs the malingerer will persist in not speaking or in not remembering, whereas the symptoms of the genuinely afflicted will temporarily disappear. Another technique is to pretend to take the deception seriously, express grave concern, and tell the "patient" that the only remedy for his illness is a series of electric shock treatments or a frontal lobotomy.

L. Conclusion

A brief summary of the foregoing may help to pull the major concepts of coercive interrogation together:

1. The principal coercive techniques are arrest, detention, the deprivation of sensory stimuli, threats and. fear, debility, pain, heightened suggestibility and hypnosis, and drugs.

2. If a coercive technique is to be used, or if two or more are to be employed jointly, they should be chosen for their effect upon the individual and carefully selected to match his personality.

3. The usual effect of coercion is regression. The interrogatee's mature defenses crumbles as he becomes more childlike. During the process of regression the subject may experience feelings of guilt, and it is usually useful to intensify these.

4. When regression has proceeded far enough so that the subject's desire to yield begins to overbalance his resistance, the interrogator should supply a face-saving rationalization. Like the coercive technique, the rationalization must be carefully chosen to fit the subject's personality.

5. The pressures of duress should be slackened or lifted after compliance has been obtained, so that the interrogatee's voluntary cooperation will not be impeded.

No mention has been made of what is frequently the last step in an interrogation conducted by a Communist service: the attempted conversion. In the Western view the goal of the questioning is information; once a sufficient degree of cooperation has been obtained to permit the interrogator access to the information he seeks, he is not ordinarily concerned with the attitudes of the source. Under some circumstances, however, this pragmatic indifference can be short-sighted. If the interrogatee remains semi hostile or remorseful after a successful interrogation has ended, less time may be required to complete his conversion (and conceivably to create an enduring asset) than might be needed to deal with his antagonism if he is merely squeezed and forgotten.

X. INTERROGATOR's CHECK LIST

The questions that follow are intended as reminders for the interrogator and his superiors.

1. Have local (federal or other) laws affecting KUBARK's conduct of a unilateral or joint interrogation been compiled and learned?

2. If the interrogatee is to be held, how long may he be legally detained?

3. Are interrogations conducted by other ODYOKE departments and agencies with foreign counterintelligence responsibilities being coordinated with KUBARK if subject to the provisions of Chief/KUBARK Directive, lor Chief/KUBARK Directive ? (b)(3) Has a planned KUBARK interrogation subject to the same provisions been appropriately coordinated?

4 Have applicable KUBARK regulations and directives been observed? These include the related Chief Directives, (b)(3) pertinent and the provisions governing duress which appear (b)(3) in various paragraphs of this handbook.

5. Is the prospective interrogatee a PBPRIME citizen? If so, have the added considerations listed on various paragraphs been duly noted?

6. Does the interrogators selected for the task meet the four criteria of (a) adequate training and experience, (b) genuine familiarity with the language to be used, (c) knowledge of the geographical cultural area concerned, and (d) psychological comprehension of the interrogatee?

7. Has the prospective interrogatee been screened? What are his major psychological characteristics? Does he belong to one of the nine major categories listed in pp. 19-28? Which?

8. Has all available and pertinent information about the subject been assembled and studied?

9. Is the source to be sent to an interrogation center, or will questioning be completed elsewhere? If at a base or station, will the

interrogator, interrogatee, and facilities be available for the time estimated as necessary to the completion of the process? If he Is to be sent to a center, has the approval of the center or of Headquarters been obtained?

10. Have all appropriate documents carried by the prospective interrogatee been subjected to technical analysis?

11. Has a check of logical overt sources been conducted? Is the interrogation necessary?

12, Have field and headquarters traces been run on the potential interrogatee and persons closely associated with him by emotional, family, or business ties?

13. Has a preliminary assessment of bona fides been carried out? With what results?

14. If an admission of prior association with one or more foreign intelligence services or Communist parties or fronts has been obtained, have full particulars been acquired and reported?

15. Has LCFLUTTER been administered? As early as practicable? More than once? When?

16. Is it estimated that the prospective interrogatee is likely to prove cooperative or recalcitrant? If resistance is expected, what is its anticipated source: fear, patriotism, personal considerations, political convictions, stubbornness, other?

17. What is the purpose of the interrogation? 18, Has an interrogation plan been prepared?

19. If the interrogation is to be conducted jointly with a liaison service, has due regard been paid to the opportunity thus, d afforded to acquire additional information about that service while minimizing KUBARK's exposure to it?

20. Is an appropriate setting for interrogation available? 21, Will the interrogation sessions be recorded? Is the equipment available? Installed?

22. Have arrangements been made to feed, bed, and guard the subject as necessary?

23. Does the interrogation plan call for more than one interrogator? If so, have roles been assigned and schedules prepared?

24. Is the interrogational environment fully subject to the interrogator's manipulation and control?

25. What disposition is planned for the interrogatee after the questioning ends?

26. Is it possible, early in the questioning, to determine the subject's personal response to the interrogator or interrogators? What is the interrogator's reaction to the subject? Is there an emotional reaction strong enough to distort results? If so, can the interrogator be replaced?

27. If the source is resistant, will non coercive or coercive techniques be used? What is the reason for the choice?

28. Has the subject been interrogated earlier? Is he sophisticated about interrogation techniques?

29. Does the impression made by the interrogatee during the opening phase of the interrogation confirm or conflict with the preliminary assessment formed before interrogation started? If there are significant differences, what are they and how do they affect the plan for the remainder of the questioning?

30. During the opening phase, have the subject's voice, eyes, mouth, gestures, silences, or other visible clues suggested areas of sensitivity? If so, on what topics?

31. Has rapport been established during the opening phase?

32. Has the opening phase been followed by a reconnaissance? What are the key areas of resistance? What tactics and how much pressure will be required to overcome the resistance? Should the ; estimated duration of interrogation be revised? If so, are further : arrangements necessary for continued detention, liaison support, guarding, or other purposes?

33. In the view of the interrogator, what is the emotional reaction of the subject to the interrogator? Why?

34, Are interrogation reports being prepared after each session, from notes or tapes?

35. What disposition of the interrogatee is to be made after questioning ends? If the subject is suspected of being a hostile agent and if interrogation has not produced confession, what measures will be taken to ensure that he is not left to operate as : before, unhindered and unchecked?

36. Are any promises made to the interrogatee unfulfilled when questioning ends? Is the subject vengeful? Likely to try to strike back? How?

37. If one or more of the non-coercive techniques discussed on pp. 52-81 have been selected for use, how do they match the subject's personality?

38. Are coercive techniques to be employed? If so, have all field personnel in the interrogator's direct chain of command been notified? Have they approved?

39. Has prior Headquarters permission been obtained?

40. Is arrest contemplated? By whom? Is the arrest fully legal? If difficulties develop, will the arresting liaison service reveal KUBARK's role or interest?

41. As above, for confinement. If the interrogatee is to be confined, can KUBARK control his environment fully? Can the normal routines be disrupted for interrogation purposes?

42. 'Is solitary confinement to be used? Why? Does the place of confinement permit the practical elimination of sensory stimuli?

43. Are threats to be employed? As part of a plan? Has the nature of the threat been matched to that of the interrogatee?

44. If hypnosis or drugs are thought necessary, has Headquarters been given enough advance notice? Has adequate allowance been made for travel time and other preliminaries?

45. Is the interrogatee suspected of malingering? If the interrogator is uncertain, are the services of an expert available?

46. At the conclusion of the interrogation, has a comprehensive

summary report been prepared?

47. Is the interrogatee to be used operationally when interrogation is over? If so, what effect (if any) is the interrogation expected to have upon the operation?

48. If the interrogation was conducted jointly with a liaison service, or was supported by liaison, how much did the host device learn about KUBARK as a result?

49. Was the interrogation a success? Why?

50. A failure? Why?

XI. DESCRIPTIVE BIBLIOGRAPHY

This bibliography is selective; most of the books and articles consulted during the preparation of this study have not been included here. Those that have no real bearing on the counter-intelligence interrogation of resistant sources have been left out. Also omitted are some sources considered elementary, inferior, or unsound. It is not claimed that what remains is comprehensive as well as selective, for the number of published works having some relevance even to the restricted subject is over a thousand. But it is believed that all the items listed here merit reading by KUBARK personnel concerned with interrogation.

1. Anonymous. Interrogation, undated. This paper is a one-hour lecture on the subject. It is thoughtful, forthright, and based on extensive experience. It deals only with interrogation following arrest and detention. Because the scope is nevertheless broad, the discussion is brisk but necessarily less than profound.

2. Barioux, Max, "A Method for the Selection, Training, and Evaluation of Interviewers," Public Opinion Quarterly, Spring 1952, Vol. 16, No. 1. This article deals with the problems of interviewers conducting public opinion polls. It is of only slight value for interrogators, although it does suggest pitfalls produced by asking questions that suggest their own answers.

3. Biderman, Albert D., A Study for Development of Improved Interrogation Techniques: Study SR 177-D (U), Secret, final report of Contract AF 18 (600) 1797, Bureau of Social Science Research Inc., Washington, D.C., March 1959. Although this book (207 pages of text) is principally concerned with lessons derived from the interrogation of American POW's by Communist services and with the problem of resisting interrogation, it also deals with the interrogation of resistant subjects. It has the added advantage of incorporating the findings and views of a number of scholars and specialists in subjects closely related to interrogation. As the frequency of citation indicates, this book was one of the most useful works consulted; few KUBARK interrogators would fail to profit from reading it. It also contains a descriminating but undescribed bibliography of 343 items.

4, Biderman, Albert D., "Communist Attempts to Elicit False

Confession from Air Force Prisoners of War", Bulletin of the New York Academy of Medicine, September 1957, Vol. 33. An excellent analysis of the psychological pressures applied by Chinese Communists to American POW's to extract "confessions" for propaganda purposes.

5. Biderman, Albert D., "Communist Techniques of Coercive Interrogation", Air Intelligence, July 1955, Vol. 8, No. 7. This short article does not discuss details. Its subject is closely related to that of item 4 above; but the focus is on interrogation rather than the elicitation of "confessions", | "

6. Biderman, Albert D., "Social Psychological Needs and | Involuntary' Behavior as Illustrated by Compliance in Interrogation", Sociometry, June 1960, Vol. 23. This interesting article is directly relevant. It provides a useful insight into the interaction between interrogator and interrogatee. It should be compared with Milton W. Horowitz's 'Psychology of Confession" (see below).

7. Biderman, Albert D. and Herbert Zimmer, The Manipulation of Human Behavior, John Wiley and Sons Inc., New York and London, 1961. This book of 304 pages consists of an introduction by the editors and seven chapters by the following specialists; Dr. Lawrence E. Hinkle Jr., "The Physiological State of the Interrogation Subject as it Affects Brain Function"; Dr. Philip E. Kubzansky, "The Effects of Reduced Environmental Stimulation on Human Behavior: A Review"; Dr. Louis A. Gottschalk, 'The Use of Drugs in Interrogation"; Dr. ' R.C. Davis, "Physiological Responses as a Means of Evaluating Information" (this chapter deals with the polygraph); Dr. Martin T. Orne, "The Potential Uses of Hypnosis in Interrogation"; Drs. Robert R. Blake ' and Jane S. Mouton, '"The Experimental Investigation of Interpersonal Influence"; and Dr. Malcolm L. Meltzer, 'Countermanipulation through Malingering." Despite the editors preliminary announcement that the book has "a particular frame of reference; the interrogation of an unwilling subject", the stress is on the listed psychological specialties; and interrogation gets comparitively short shrift. Nevertheless, the KUBARK interrogator should read this book, especially the chapters by Drs. Orne and Meltzer, He will find that the book is by scientists for scientists and that the contributions consistently demonstrate too theoretical an understanding of interrogation per se. He will also find that practically no valid experimentation the results of which were unclassified and available to the authors has been conducted under interrogation conditions. Conclusions are

suggested, almost invariably, on a basis of extrapolation. But the book does contain much useful information, as frequent references in this study show. The combined bibliographies contain a total of 771 items. '

8. — A good, brief discussion of the purpose, tools, and techniques employed in the interrogation of arrestees. Although the author says that his essay "is slanted toward relatively unsophisticated cases, and does / not cover the subtler techniques...", he manages in a very short paper to discuss a number of the essentials of questioning resistant sources. Interrogators will find that much of the material is familiar but that the article makes rewarding reading nonetheless.

9. - All interrogators should read this short, authoritative essay.

10. - This article is a review of current hypotheses about the reliability of information obtained from a subject in trance, the hypnosis of unwilling subjects, attempts to induce the performance of crimes through hypnosis, and the possible prophylactic value of hypnosis as a defense against interrogation. The author obviously speaks with a good deal of authority. Most of his conclusions are negative-i.e., hypnosis can be a useful aid for interrogators but is far from a magic solution for all problems.

11. Farber, I. E., Harry F. Harlow, and Louis Jolyon West, "Brainwashing, Conditioning, and DDD," Soctometry, December 1957, Vol. 20, No. 4. The "DDD" refers to the debility, dependency, and dread syndrome, postulated by the authors are the three essentials of the "brainwashing" process. The article is well worth reading.

12. This article provides some sound information but the discussion of interrogation as such, though clear and are well-ordered, contains a few questionable postulates. The article merits reading but is not recommended as a guide to the conduct of interrogation.

13. Gill, Merton, Inc., and Margaret Brenman, Hypnosis and Related States: Psychoanalytic Studies in Regression, International Universities Press Inc., New York, 1959. This book is a scholarly and comprehensive examination of hypnosis. The approach is basically Freudian but the authors are neither narrow nor doctrinaire. The book discusses the induction of hypnosis, the hypnotic state, theories of induction and of the hypnotic condition, the concept of regression as a basic element in hypnosis, relationships between

hypnosis and drugs, sleep, fugue, etc., and the use of hypnosis in psychotherapy. Interrogators may find the comparison between hypnosis and "brainwashing" in chapter 9 more relevant than other parts. The book is recommended, however, not because it contains any discussion of the employment of hypnosis in interrogation (it does not) but because it provides the interrogator with sound information about what hypnosis can and cannot do.

14. Hinkle, Lawrence E. Jr. and Harold G. Wolff, "Communist Interrogation and Indoctrination of Enemies ' of the State", AMA Archives of Neurology and Psychiatry, August 1956, Vol. 76, No, 2. This article summarizes the physiological and psychological reactions of American prisoners to Communist detention and interrogation, It merits reading but not study, chiefly because of the vast differences between Communist interrogation of American POW's and KUBARK interrogation of known or suspected personnel of Communist services or parties.

15. Horowitz, Milton W., "Psychology of Confession. " Journal of Criminal Law, Criminology, and Police Science, JulyAugust 1956, Vol. 47. The author lists the following principles of confession: (1) the subject feels accused; (2) he is confronted by authority wielding power greater than his own; (3) he believes that evidence damaging to him is available to or possessed by the authority: (4) the accused is cut off from friendly support; (5) self-hostility is generated; and (6) confession to authority promises relief. Although the article is essentially a speculation rather than a report of verified facts, it merits close reading.

16. Inbau, Fred E, and John E, Reid, Lie Detection and Criminal Investigation, Williams and Wilkins Co., 1953. The first part of this book consists of a discussion of the polygraph. It will be more useful to the KUBARK interrogator than the second, which deals with the elements of criminal interrogation.

17. KHOKHLOV, Nicolai, In the Name of Conscience, David McKay Co., New York, 1959. This entry is included chiefly because of the cited quotation. It does provide, however, some interesting insights into the attitudes of an interrogatee.

18. KUBARK, Communist Control Methods, Appendix 1: "The Use of Scientific Design and Guidance Drugs and Hypnosis in Communist Interrogation and Indoctrination Procedures." Secret, no date, The appendix reports a study of whether Communist

interrogation methods included such aids as hypnosis and drugs. Although experimentation in these areas is, of course, conducted in Communist countries, the study found no evidence that such methods are used in Communist interrogations -- or that they would be necessary.

19. KUBARK (KUSODA), Communist Control Techniques, Secret, 2 April 1956. This study is an analysis of the methods used by Communist State police in the arrest, interrogation, and indoctrination of persons regarded as enemies of the state. This paper, like others which deal with Communist interrogation techniques, may be useful to any KUBARK interrogator charged with questioning a former member of an Orbit intelligence or security service but does not deal with interrogation conducted without police powers.

20. KUBARK, Hostile Contro] and Interrogation Techniques, Secret, undated, This paper consists of 28 pages and two annexes. It provides counsel to. KUBARK personnel on how to resist inte rrogation conducted by a hostile service. Although it includes sensible advice on resistance, it does not present any new information about the theories or practices of interrogation. rc

21.- 22. (Deleted)

23. Laycock, Keith, 'Handwriting Analysis as an Assessment Aid, " Studies in Intelligence, Summer 1959, Vol. 3, No. 3. A defense of graphology by an "educated amateur, "' Although the article i's interesting, it does not present tested evidence that the analysis of a subject's handwriting would be a useful aid to an interrogator. Recommended, nevertheless, for interrogators unfamiliar with the subject.

24. Lefton, Robert Jay, "Chinese Communist 'Thought Reform.': Confession and Reeducation of Western Civilians, "' Bulletin of the New York Academy of Medicine, September 1957, (Vol. 33. A sound article about Chicom brainwashing techniques. The information was compiled from first-hand interviews with prisoners who had been subjected to the process, Recommended as background reading.

25. Levenson, Bernard and Lee Wiggins, A Guide for Intelligence Interviewing of Voluntary Foreign Sources, Official Use Only, Officer Education Research Laboratory, ARDC, Maxwell Air Force Base (Technical Memorandum OERL-TM-54-4.) A good, though

generalized, treatise on interviewing techniques. As the title shows, the subject is different from that of the present study.

26. Lilly, John C,, "Mental Effects of Reduction of Ordinary Levels of Physical Stimuli on Intact Healthy Persons. ' Psychological Research Report #5, American Psychiatric Association, 1956. After presenting a short summary of a few autobiographical accounts written about relative isolation at sea (in small boats) or polar regions, the author describes two experiments designed to mask or drastically reduce most sensory stimulation. The effect was to speed up the results of the more usual sort of isolation (for example, solitary confinement). Delusions and hallucinations, preceded by other symptoms, appeared after short periods. The author does not discuss the possible relevance of his findings to interrogation.

27. Meerlo, Joost A.M., The Rape of the Mind, World Publishing Co., Cleveland, 1956, This book's primary value for the interrogator is that it will make him aware of a number of elements in the responses of an interrogatee which are not directly related to the questions asked or the interrogation setting but are instead the product of (or are at least influenced by) all questioning that the subject . has undergone earlier, especially as a child, For many interrogatees the interrogator becomes, for better or worse, the parent or authority symbol. Whether the subject is submissive or belligerent may be determined in part by his childhood relationships with his parents. Because the same forces are at work in the interrogator, the interrogation may be chiefly a cover for a deeper layer of exchange or conflict between the two. For the interrogator a primary value of this book (and of much related psychological and psychoanalytic work) is that it may give him a deeper insight into himself.

28. Moloney, James Clark, "Psychic Self-Abandon and Extortion of Confessions, "' International Journal of Psychoanalysis, January/ February 1955, Vol. 36, This short article relates the psychological release obtained through confession (i.e., the sense of well-being following surrender as a solution to an otherwise unsolvable conflict) with religious experience generally and some ten Buddhistic practices particularly. The interrogator will find little here that is not more helpfully discussed in other sources, including Gill and Brenman's Hypnosis and Related States. Marginal.

29. Oatis, William N., "Why I Confessed, " Life, 21 September 1953, Vol. 35. Of some marginal value because it combines the

writer's profession of innocence ("I am not a spy and never was") with an account of how he was brought to "'confess'" to espionage within three days of his arrest. Although Oatis was periodically deprived of sleep (once for 42 hours) and forced to stand until weary, the Czechs obtained the "confession" without torture or starvation and without sophisticated techniques.

30. Rundquist, E,A., "The Assessment of Graphology," Studies in Intelligence, Secret, Summer 1959, Vol. 3, No. 3. The author concludes that scientific testing of graphology is needed to permit an objective assessment of the claims made in its behalf. This article should be read in conjunction with No, 23, above.

31. Schachter, Stanley, The Psychology of Affiliation: Experimental Studies of the Sources of Gregariousness, Stanford University Press, Stanford, California, 1959. A report of 133 pages, chiefly concerned with experiments and statistical analyses performed at the University of Minnesota by Dr. Schachter and colleagues. The principal findings concern relationships among anxiety, strength of affiliative tendencies, and the ordinal position (i.e., rank in birth sequence among siblings). Some tentative conclusions of significance for interrogators are reached, the following among them:
a. "One of the consequences of isolation appears to be a psychological state which in its extreme form resembles a full-blown anxiety attack." (p. 12.)
b. Anxiety increases the desire to be with others who share the same fear.
c. Persons who are first-born or only children are typically more nervous or afraid than those born later. Firstborns and onlies are also "considerably less willing or able to withstand pain than are later-born children." (p. 49.)
In brief, this book presents hypotheses of interest to interrogators, but much further research is needed to test validity and applicability.

32. Sheehan, Robert, Police Interview and Interrogations and the Preparation and Signing of Statements. A 23-page pamphlet, unclassified and undated, that discusses some techniques and tricks that can be used in counterintelligence interrogation. The style is sprightly, but most of the material is only slightly related to KUBARK's interrogation problems. Recommended as background reading.

33, Singer, Margaret Thaler and Edgar H. Schein, "Projective

Test Responses of Prisoners of War Following Repatriation." Psychiatry, 1958, Vol. 21. Tests conducted on American ex-POW's returned during the Big and Little Switches in Korea showed differences in characteristics between non-collaborators and collaborators., The latter showed more typical and humanly responsive reactions to psychological testing than the former, who tended to be more apathetic and emotionally barren or withdrawn. Active resisters, however, often showed a pattern of reaction or responsiveness like that of collaborators. Rorschach tests provided clues, with a good statistical incidence of reliability, for differentiation between collaborators and non-collaborators. The tests and results described are worth noting in conjunction with the screening procedures recommended in this paper.

34. Sullivan, Harry Stack, The Psychiatric Interview, W.W. Norton and Co,, New York, 1954. Any interrogator reading this book will be struck by parallels between the psychiatric interview and the interrogation, The book is also valuable because the author, a psychiatrist of considerable repute, obviously had a deep understanding of the nature of the inter-personal relationship and of resistance.

35. U.S, Army, Office of the Chief of Military History, . Russian Methods of Interrogating Captured Personnel in World War II, Secret, Washington, 1951. A comprehensive treatise on Russian intelligence and police systems and on the history of Russian treatment of captives, military and civilian, during and following World War Il. The appendix contains some specific case summaries of physical torture by the secret police. Only a small part of the book deals with interrogation, Background reading.

36. U.S, Army, 7707 European Command Intelligence Center, Guide for Intelligence Interrogators of Eastern Cases, Secret, April 1958. This specialized study is of some marginal value for KUBARK interrogators dealing with Russians and other Slavs.

37. U.S. Army, The Army Intelligence School, Fort Holabird, Techniques of Interrogation, Instructors Folder I-6437/A, January 1956. This folder consists largely of an article, "Without Torture, "' by a German ex-interrogator, Hans Joachim Scharff. Both the preliminary discussion and the Scharff article (first published in Argosy, May 1950) are exclusively concerned with the interrogation of POW's. Although Scharff claims that the methods used by German Military Intelligence against captured U.S. Air Force

personnel". . . were almost irresistible,'" the basic technique consisted of impressing upon the prisoner the false conviction that his information was already known to the Germans in full detail, The success of this method depends upon circumstances that are usually lacking in the peacetime interrogation of a staff or agent member of a hostile intelligence service. The article merits reading, nevertheless, because it shows vividly the advantages that result from good planning and organization.

38. U.S. Army, Counterintelligence Corps, Fort Holabird, Interrogations, Restricted, 5 September 1952. Basic coverage of military interrogation. Among the subjects discussed are the interrogation of witnesses, suspects, POW's, and refugees, and the employment of interpreters and of the polygraph, Although this text does not concentrate upon the basic problems confronting KUBARK interrogators, it will repay reading.

39. U.S, Army, Counterintelligence Corps, Fort Holabird, Investigative Subjects Department, Interrogations, Restricted, 1 May 1950. This 70-page booklet on counterintelligence interrogation is basic, succinct, practical, and sound. Recommended for close ' reading.

40. U.S, Defector Reception Center, Defector Reception Center Procedures Manual, Secret, 1 January 1956. Almost wholly devoted to the administration and handling of defectors and refugees, the manual devotes only two generalized pages to interrogation. KUBARK personnel concerned with reception center processing should read it.

41. Wellman, Francis L., The Art of Cross-Examination, Garden City Publishing Co. (now Doubleday), New York, originally 1903, 4th edition, 1948, Most of this book is but indirectly related to the subject of this study; it is primarily concerned with tripping up witnesses and impressing juries. Chapter VIII, "Fallacies of Testimony, '" is worth reading, however, because some of its warnings are applicable.

42. Wexler, Donald, Jack Mendelson, Herbert Leiderman, and Philip Solomon, "Sensory Deprivation, '" A,.M,A, Archives of Neurology and Psychiatry, 1958, 79, pp. 225-233. This article reports an experiment designed to test the results of eliminating most sensory stimuli and masking others. Paid volunteers spent periods from 1 hour and 38 minutes to 36 hours in a tank-respirator, The

results included inability to concentrate effectively, daydreaming and fantasy, illusions, delusions, and hallucinations, The suitability of this procedure as a means of speeding up the effects of solitary confinement upon recalcitrant subjects has not been considered.

OTHER BIBLIOGRAPHIES

The following bibliographies on interrogation were noted during the preparation of this study.

1. Brainwashing, A Guide to the Literature, prepared by the Society for the Investigation of Human Ecology, Inc., Forest Hills, New York, December 1960. A wide variety of materials is represented: scholarly and scientific reports, governmental and organizational reports, legal discussions, biographical accounts, fiction, journalism, and miscellaneous. The number of items in each category is, respectively, 139, 28, 7, 75, 10, 14, and 19, a total of 418. One or two sentence descriptions follow the titles. These are restricted to an indication of content and do not express value judgements. The first section contains a number of especially useful references.

2. Comprehensive Bibliography of Interrogation Techniques, Procedures, and Experiences, Air Intelligence Information Report, Unclassified, 10 June 1959. This bibliography of 158 items dating between 1915 and 1957 comprises "the monographs on this subject available in the Library of Congress and arranged in alphabetical order by author, or in the absence of an author, by title.'" No descriptions are included, except for explanatory sub-titles. The monographs, in several languages, are not categorized. This collection is extremely heterogeneous. Most of the items are of scant or peripheral value to the interrogator.

3. Interrogation Methods and Techniques, KUPALM, L-3,024,941, July 1959, Secret/NOFORN. This bibliography of 114 items includes references to four categories: books and pamphlets, articles from periodicals, classified documents, and materials from classified periodicals. No descriptions (except sub-titles) are included. The range is broad, so that a number of nearly-irrelevant titles are included (e.g., Employment psychology: the Interview, Interviewing in social research, and "Phrasing questions; the question of bias in interviewing", from Journal of Marketing).

4. Survey of the Literature on Interrogation Techniques, KUSODA, 1) March 1957, Confidential. Although now somewhat dated because of the significant work done since its publication, this bibliography remains the best of those listed. It groups its 114 items in four categories: Basic Recommended Reading, Recommended Reading, Reading of Limited or Marginal Value, and Reading of

No Value. A brief description of each item is included. Although some element of subjectivity inevitably tinges these brief, critical appraisals, they are judicious; and they are also real time-savers for interrogators too busy to plough through the acres of print on the specialty.

Glossary

KUBARK: Central Intelligence Agency (CIA)

LCFLUTTER: Polygraph, sometimes supplanted by truth drugs: Sodium Amytal (amobarbital), Sodium Pentothal (thiopental), and Seconal (secobarbital) to induce regression in the subject

KUCAGE: CIA Psychological and Paramilitary Operations Staff

KUDOVE: CIA Deputy Director for Operations (DDO)

KUPALM: CIA Office of Central Reference

NOFORN: No foreign nationals

ODYOKE: Federal government of the United States

AVH: Hungarian State Security

KGB: Soviet State Security

PBPRIME: United States

ODENVY: Federal Bureau of Investigation

KUDESK: CIA Counterintelligence Center

CI: Counterintelligence

www.ingramcontent.com/pod-product-compliance
Lightning Source LLC
Chambersburg PA
CBHW050540280326
41933CB00011B/1662